Contemporary Trends and Issues in Science Education

Volume 56

The book series Contemporary Trends and Issues in Science Education provides a forum for innovative trends and issues impacting science education. Scholarship that focuses on advancing new visions, understanding, and is at the forefront of the field is found in this series. Authoritative works based on empirical research and/or conceptual theory from disciplines including historical, philosophical, psychological and sociological traditions are represented here. Our goal is to advance the field of science education by testing and pushing the prevailing sociocultural norms about teaching, learning, research and policy. Book proposals for this series may be submitted to the Publishing Editor: Claudia Acuna E-mail: Claudia.Acuna@springer.com

Ben Akpan • Bulent Cavas • Teresa Kennedy
Editors

Contemporary Issues in Science and Technology Education

 Springer

Editors
Ben Akpan
The STAN Place
Science Teachers Association of Nigeria
Abuja, Nigeria

Bulent Cavas
Faculty of Education
Dokuz Eylul Universitesi
Buca, Turkey

Teresa Kennedy
SOE, BEP 229F
University of Texas at Tyler
Tyler, TX, USA

ISSN 1878-0482   ISSN 1878-0784   (electronic)
Contemporary Trends and Issues in Science Education
ISBN 978-3-031-24258-8   ISBN 978-3-031-24259-5   (eBook)
https://doi.org/10.1007/978-3-031-24259-5

This Springer imprint is published by the registered company Springer Nature Switzerland AG
The registered company address is: Gewerbestrasse 11, 6330 Cham, Switzerland

Foreword

I feel highly honoured by the International Council of Associations for Science Education (ICASE) for asking me to write the *Foreword* for its 50th anniversary commemorative book, *Contemporary Issues in Science and Technology Education*. It is even more heart-warming that the 50th anniversary of ICASE is being marked at its 2023 world conference, which the British University in Dubai (BUiD) is extremely privileged to host jointly with our partner, Amity University Dubai. I would like to congratulate ICASE on its 50th anniversary and for its immense contribution to the growth and development of science and technology education globally over the past five decades.

The contents of this book have been carefully selected to match the title of the book. The world is experiencing interesting times and facing challenging issues. These include climate change, food shortages and cross-boundary health issues. Science and technology throughout history have led the development of civilisations and societies. Science and technology solutions have protected humankind from destruction and have contributed to the creation of the prosperity and the healthy style of living that many societies enjoy today. I have no doubt that science and technology will continue to lead the way forward.

Science and technology education currently faces various challenges. The world in which we live today is different from that of our ancestors. The current issues are more complex. The gap between developed and developing countries is expanding. Hence, scientists alone will not be able to solve the global issues. Partnerships throughout society are essential. Science and technology education professionals should collaborate more innovatively with the public and private sectors. More entrepreneurship activities should be encouraged. In my view, science and technology education delivery requires a major overhaul. Stakeholders would need to go to a clean whiteboard to redesign curricula, and to re-consider the methods of instruction, assessment and evaluation. In this new world of digital technologies, we cannot continue to do what we have so far been doing. There is an overarching need to redefine and optimise the role of humans in relation to what machine and digital technologies can do for us. More interaction with other disciplines should be initiated. Indeed, significant dialogue needs to start between ICASE and similar global professional organisations towards promoting academic interdisciplinary research, critical thinking and open forums for knowledge exchange. Funding projects that have a significant impact should be given priority and governments should allocate funds for these.

This book should be seen in light of the foregoing. Its publication is, therefore, not only timely but a step in the right direction. With great delight, I commend its use by policymakers, researchers, undergraduate and post-graduate students in science and technology education, as well as their teachers.

Vice Chancellor, The British University in Dubai   Abdullah M. Alshamsi
Dubai, United Arab Emirates

Contents

List of Figures

List of Tables

Chapter 1
Introduction – A Reflection on Contemporary Issues in Science and Technology Education

Ben Akpan

Abstract This chapter provides an introduction to the book *Contemporary Issues in Science and Technology Education*. It discusses the three broad themes under which the various chapters are grouped. The themes are: philosophical foundations and curriculum development; sustainable development, technology and society; and the learning sciences and twenty-first century skills. The purpose of this introductory chapter is to give the reader an insight into the issues that are discussed in the various chapters in the book.

Keywords Philosophical foundations · Curriculum development · Curriculum implementation · Sustainable development · Science/technology/society · The learning sciences · Twenty-first century skills

Introduction

The term *contemporary* may be used to refer to some things occurring at the same time; something occurring in the present; or one thing that exists at the same time as another. For the purposes of this book, the second meaning – *something occurring in the present* – is the most relevant. Similar terms include *present day* and *current*. Contemporary issues in science and technology education (STE) therefore refer to ideas, opinions, and/or topics that are encountered in present day teaching and learning of science and technology (S&T) subjects. This book, *Contemporary Issues in Science and Technology Education*, discusses some of the current issues in STE. These issues have been grouped into three broad themes: philosophical foundations and curriculum development; sustainable development, technology and society; and the learning sciences and twenty-first century skills. The grouping is by no means an attempt to define rigid boundaries but rather a guide to fashioning out

B. Akpan (✉)
The STAN Place, Abuja, Nigeria

© The Author(s), under exclusive license to Springer Nature Switzerland AG 2023
B. Akpan et al. (eds.), *Contemporary Issues in Science and Technology Education*, Contemporary Trends and Issues in Science Education 56, https://doi.org/10.1007/978-3-031-24259-5_1

1

the orientation of the book. Some chapters could very much fit into different groups. Some groupings illustrate synergy rather than disparity. For instance, philosophical foundations and curriculum development are in one group to demonstrate the idea that curriculum and teaching should be built upon sound philosophical foundations. The object of this introductory chapter is to provide an overview of these broad themes in order to give the reader an insight into the issues raised in the various chapters.

Philosophical Foundations and Curriculum Development

Philosophy is the study of the fundamental nature of knowledge, reality, and existence. In S&T, philosophy concerns itself with how we come to know about phenomena as well as the processes and values by which S&T build a coherent body of knowledge, because philosophical foundations of S&T differ from other philosophical views. S&T obtain their knowledge empirically based on observation, either directly or indirectly through the use of apparatus. Much of the effort of personnel in S&T is actually devoted to observing and measuring phenomena more accurately in order to obtain better empirical data. Logic and reason are naturally involved in these efforts, but S&T specialists do not accept knowledge based only on the basis of logic unless they can be verified by some empirical means. The empirical approach in S&T is based on the assumptions that: (1) the universe is intelligible; (2) we can study nature and discover natural laws; (3) there is reality within space, time and matter; and (4) all natural phenomena can be explained in terms of physical and chemical states.

In S&T, explanations are not made in teleological terms, but in mechanistic terms. *Teleology* is the doctrine of *final causes* – explaining phenomena in terms of the purpose that they serve rather than of the cause by which they arise. Theories in S&T are based on facts that are derived from observation and experimentation. If further experimentation reveals new information, theories have to be modified. Since theories evolve and are modified as knowledge of phenomena increases, the goal of S&T in formulating broad, encompassing theories never ends. So, S&T personnel are humble about what they know. At their core, therefore, S&T are ways of properly understanding nature that are full of regularities. Seen in this context, S&T have been contemporaneous with *Homo Sapiens* as a species. Fortunately, as is often said, it is not just wonderful that humanity understands nature, but also that the universe is understandable. Even so, the knowledge of the universe through S&T is extremely limited by the capacity of our senses of sight, touch, and hearing, as well as of the instruments manufactured by humankind. Indeed, it is now known that scientific knowledge preceded writing, while astronomy woven around theology received great attention in the BCE (*Before the Christian Era*) years. Thales of Miletus (c.625–545 BC) was the first natural philosopher of note. Thales explained phenomena in nature based on the changes in the state of water: solid, liquid and gaseous states. Teleology was the dominant explanation of phenomena – objects in the universe had purposes and they moved naturally towards those destined ends. Then came Pythagoras, c.582–500 BC, a native of Samos (a Greek Island in the east Aegean, just off the coast of modern-day Turkey), who founded a brotherhood devoted to a life of mathematical speculation and religious contemplation. For the Pythagoreans, numbers provided a conceptual model of the universe, quantities and shapes determining the forms of natural objects. Socrates (470–399 BC) would later consider astronomy a waste of time, stating instead that the prime task of the philosopher was the ordering of humans in society, not the understanding or the control of nature. The work of Socrates was continued by his pupil, Plato (427–347 BC), who saw that any philosophy with a claim to generality must include a theory as to the nature of the universe. Plato removed the taint of atheism from astronomical studies and maintained that natural laws were subordinate to the authority of divine principles. For example, events occur primarily because rational purposes and designs are formulated by intelligent beings. The works of Aristotle (384–322 BC) marked a turning point in the history of Greek science, being the last to formulate a system to the world as a whole, and he was the first to embark upon extensive empirical enquiries. Observational astronomy revived in the fifteenth century, such that a considerable body of accurate observations was available when Nicolas Copernicus (1473–1543) began his work. In the Copernican system, the Earth revolved round the Sun, in the same way as the other planets, thus bringing in an entirely new set of cosmic values. The Sun at the centre of the universe was the governor of the heavens. More observational works continued with Johann Kepler, Tycho Brache, William Gilbert and Francis Bacon, before Galileo Galilei (1564–1642) came on the scene. With his unprecedentedly powerful telescopes, Galileo was the first to observe the uneven, cratered surface of the Moon; Jupiter's four largest satellites; dark spots on the surface of the Sun; and the phases of Venus. With Galileo, the mathematical experimental method of science came to maturity. According to his principle of *inertia*, if the surface of the Earth were perfectly smooth, a sphere set in motion on that surface would continue to roll round the Earth indefinitely. So, Galileo thought that uniform speed in a circle was the natural motion of all bodies that were not acted upon by a force. Being constrained by one ancient conception, which he never overcame, namely the idea that motions of the heavenly bodies were circular and uniform, Galileo failed at this point. However, the problems emerging from the Galilean era were addressed subsequently by other scientists, including Isaac Newton (1643–1727). In optics, Newton's discovery of the composition of white light integrated the phenomenon of colours into the science of light and laid the foundation for physical optics. In mechanics, his three laws of motion resulted in the formulation of the law of universal gravitation. In mathematics, Newton was the original discoverer of the infinitesimal calculus. Newton's *Philosophiae Naturalis Principia Mathematica (Mathematical Principles of Natural Philosophy)* (1687) is one of the most important single works in the history of modern science. Isaac Newton's studies would serve as a fulcrum for future works, including those of Albert Einstein (1879–1955). Einstein is, without doubt, famous for his theory of relativity, which revolutionised our understanding of space, time, gravity and the universe. Relativity also showed us that matter and energy are two different forms of the same thing – a fact that Einstein expressed as $E = mc^2$, the most famous equation in physics. However, relativity is only one part of Einstein's legacy. He was equally inventive when it came to the physics of atoms, molecules and light and, today, technological products from his works include paper towels, stock market forecasts, solar power and laser printers. Einstein also made important contributions to the development of the theory of quantum mechanics. It is interesting to compare the viewpoints of Albert Einstein with those of Isaac Newton in respect of *gravity*. According to Newton, gravity is a force with an unknown mechanism that moves at infinite speed and is tied to mass. For him, space and time are two distinct entities – absolute and fixed. But Einstein maintained that gravity is geometry and that its mechanism is curvature. In his view, gravity acts locally, moves at light speed, and is tied to energy. For Einstein, spacetime is one united entity – relative to the observer and dynamic. Chapter 2 and, to a large extent, Chap. 3 provide perspectives on the philosophy and nature of scientific and technological ideas.

In STE, the process of curriculum development has a three-way orientation: towards society, towards the learner, and towards the particular scientific knowledge that it is the responsibility of the educational institution to pass on. Thus, at a time of globalisation such as we are currently witnessing, it is necessary to understand the nature of society as it is and to extrapolate likely future trends in order to provide an adequate curriculum. To a large extent, then, the curriculum will be shaped by the culture of the society in which it operates. It will be affected by social values, social needs and social problems. If the curriculum remains static in a dynamic society, especially in periods of rapid social change, it is likely that the education that is meant to induct the young into society, and to promote an intelligent understanding of it, will cater only for needs and values that no longer exist. Indeed, it may be further lacking in that new needs have arisen for which no attempt is made to cater. In order to deal with problems such as these, it is obvious that knowledge of society is required, that there should be a careful and detailed analysis of changes and trends in the other social institutions. This awareness of society and its needs and problems will probably be dual in nature, those concerned with facts and those concerned with values. Efforts, therefore, will necessarily be directed not only towards things as they are and might be – that is, with descriptive subject matter – but also with the normative, with the idea of what society ought to be and with the possible impact of this on the educational institution. STE has a *dual mandate*, namely a preparation for S&T literacy for all and a preparation for those seeking higher careers in S&T professions. So, the end product of STE is the modification of the learner's behaviour in pursuit of these mandates. It is clear, therefore, that another aspect that should be taken into account when framing S&T curricula is the nature of the learners – their growth needs, the sequences of their development, their experiences, interests, motives and aspirations; and their relationships within their psychological life-space. This all implies that, as well as knowing the learner, we must know how the learning process can be most effectively set in motion and used, and how they are affected by the types of situations in which learning can occur. In addition to a knowledge of the nature of the individual learner and the nature of the society, there should also be a concern with the learning process and the subject matter on which it is exercised. One curricular problem is how to achieve a good balance between the acquisition of specific skills and the production of general understanding. Essentially, therefore, it is the problem of educational *transfer* – the extent to which mastery of a particular learning task produces more or less limited gains in mastery of other activities. The source material for S&T curricula is the disciplines that constitute contemporary knowledge and enable us to function in our environment. However, it is not enough merely to identify the subject matter of STE. The relations between the disciplines must be considered in order to determine what may or may not be appropriate and what decisions should be made about the sequence of instruction. In S&T, curriculum development and implementation comprise the following stages: (1) identification of goals and objectives; (2) selection of learning experiences in pursuit of the goals and objectives; (3) selection of subject matter content; (4) organisation and integration of learning experiences; and (5) evaluation of the effectiveness of the curriculum in attaining the set goals and objectives. The stages are related and interdependent and combine to form a cyclical process. Each stage is, therefore, a logical development from the preceding one. The overall goal, of course, is the change in behaviour of the learners upon exposure to the curriculum. There are a number of factors that influence changes in behaviour that have been identified by psychology. These are: (1) *individual differences* – a curriculum that sets out to change behaviour must take some account of the fact that changes occur differently in different individuals; (2) *motivation* – no change in behaviour can occur without the presence of motivation, and the problem is to identify what types of motivation will serve to bring about the particular types of changes in behaviour that may be contemplated by the builders of a curriculum in S&T, and how these kinds of motivation may be established in the learners; and (3) *learning* – how the content of a curriculum, its topics and their sequence, affect the learning of the anticipated behaviour. Chapters 4, 5, 6, 7, 8, and 9 provide insights into curriculum issues of our time.

Sustainable Development, Technology and Society

A major concern in STE is that we may be producing a generation that is out of touch with the world and its surroundings. There is a serious concern that, while our generation thinks that it is one of the most educated and informed, there are many people out there who don't understand the concept of *saving* the Earth. They don't care about *sustainable development* – a term that refers to the attainment of a balance between environmental protection and human economic development, and between present and future needs. Sustainable development involves three important aspects: economic development, social development, and environmental protection. It is a concept that is largely connected with the basic needs of individuals and requires an integration of economic, social and environmental approaches for implementation. Unless we understand the interdependency of the environment in which we live, our planet is doomed. We can decide our fate with regard to the environment, but only if we fully understand the consequences of our actions. Essentially, we must be able to supply sufficient food, energy, raw material and any required manufactured products to our citizens and other nations without compromising the world's resources for future generations, and without leaving a barren wasteland of environmental degradation. The four principal components of the ecosphere under threat are: (1) *the climate system:* a highly complex system consisting of three major components – the atmosphere, the hydrosphere, the cryosphere, and the interactions between them; (2) *the nutrient cycle* – a system whereby energy and matter are transferred between living organisms and non-living parts of the environment; (3) *the hydrological cycle* – a cycle that involves the continuous circulation of water in the Earth's atmosphere through evaporation, transpiration, condensation, precipitation and runoff; and (4) *biodiversity* – the enormous variety of life on Earth. Population growth is another major cause of concern due to the stress that it imposes on the environment. So, also, are the huge differences in wealth, life chances, health, education and the provision of social amenities that exist in various countries. Other problems lie in the area of agriculture and are caused by over-intensive and inappropriate land use, the clearing of vital vegetation, salinisation, laterisation and pollution from fertilisers and pesticides such as nitrates. Efficient cultivation, such as crop rotation, and less emphasis on the use of environmentally harmful fertilisers and pesticides are some of the methods to mitigate these effects. The use of solar, wind, wave, tide and biomass energy should be made as alternatives to conventional fossil fuels. This requires concerted global effort. There is a clear need for the proper management of the Earth's resources, which should aim to maximise the chances of achieving global sustainable development, reducing global pollution and poverty, and increasing the life chances of individuals in their various countries. Education will largely constitute a major plank in such an endeavour. It is not too late for adults to change their ways – they truly can. But soon, it will be up to the children who we are educating today to put the brakes on shortsighted environmental destruction. This is why S&T education should leverage on environmental education to assure success in this direction. Children should be excited about getting a more secure future for themselves. Indeed, it is of overarching importance for people to understand the connection between their daily lives and the environmental tragedies that they read about in social and mainstream media. Too often we forget just how intertwined our lives have become with the major forms of pollution in the world. We forget that our daily dependence on such things as cars and plastics fuels the industry that we are so quick to blame. The only way that changes are going to be made is when people understand that, if they want cleaner air or less garbage on their beaches, they are going to have to make changes at home and in their mindset. It is unfortunate that we find ourselves locked in a battle between humankind and nature. And, more unfortunately, at this moment, *nature is losing ground fast*. What we need now, and in the immediate future, are people motivated by the goal of stopping the destruction, pushing to end this drift, through local, national or international networks. It is a difficult challenge and our focus must be on solutions that will ensure that we *save the only world we know*.

Technology has impacted on many areas of our lives and has virtually brought the world to our fingertips. Several technological products have been developed, which have continued to assist people at home and in offices. Technology has made information more accessible, improving communication as well as transportation. Mobile phones and the Internet have facilitated faster and very efficient communication. New and emerging technologies are impacting our daily lives in every field. Cloud computing has improved data security and encryption. Programmers, database managers, hardware engineers and network analysts are enjoying improved and more efficient working environments. There is no doubt that advances in S&T have made life much easier. Works can be accomplished more easily through high-tech machines and equipment. The computer has become a basic companion of professionals in all cadres and vocations. Business and financial transactions can now be done relatively seamlessly across the world. Cloud-based video communication apps such as *Zoom* have facilitated virtual 'face-to-face' activities in education, training and business. Medical technology has come to the rescue, with the vaccines to combat COVID-19 developed and manufactured in record time. And, on 25th December 2021, the world was stunned with the launch of the latest and best telescope in history – the *James Webb Space Telescope* (JWST). JWST will be able to look back in time about 100 million years after the *Big Bang*, when stars and planets were beginning to form. With so much achievement in S&T, there are obvious challenges for S&T educators. STE must keep pace with developments in the larger world as well as in the world of work. Humanity expects no less. Four chapters, all in Part II of this book, highlight major issues that relate to sustainable development with links to technology and society.

The Learning Sciences and Twenty-First Century Skills

The learning sciences are an interdisciplinary group of subjects comprising cognitive science, psychology, neuroscience, computer science, machine learning, statistics, psychometrics, linguistics, and data analytics that are known to facilitate learning. This group has come about because researchers from the various fields have been examining the prospects of new ways to think about and develop educational strategies that have the backing of empirical research. It is therefore in the nature of learning sciences to conduct research on the process of learning in settings that are realistic. These settings include homes, schools, museums and other out-of-school environments. Fusco (2020) maintains that learning sciences research has the following characteristics: (1) explores beyond general principles of learning by focusing on concepts that are very important and by looking for evidence of learning in ways that traditional tests are incapable of measuring; (2) designs approaches that are innovative to learning and assessment and which are anchored on the creative use of technology and social, collaborative perspectives of learning; and (3) studies learning activity systems as a whole instead of as separate components – such systems integrate leadership management, teacher professional development, teaching resources, technologies and assessment procedures. The learning sciences are also characterised by their orientation to address equity, ensure that learners are empowered and that successful efforts are given premium attention. By valuing diversity, the field seeks to partner with people and organisations so as to understand their particular contexts and thus explore the process of learning in various places. According to Papa (2020), there are two major goals of the learning sciences. The first goal is to fashion *models of mind* in the form of theories, principles and practices through which to produce learning environments that promote problem-solving skills. This goal is exemplified by the works of Benjamin Bloom and Robert Gagné. Bloom viewed the mind as a hierarchically ordered phenomenon, which was progressively evolving by acquiring and recalling information, understanding the potential of the acquired information, applying the acquired information to solve problems, and analysing/creating/sharing the knowledge and skills that facilitate problem-solving. For his part, Gagné provided the following parameters that should lead learners to gradually but effectively apply, analyse and evaluate any acquired information: discriminations, concrete concepts, defined concepts, rules, and higher order rules. For Gagné, *meta-cognitive skills* (see Chaps. 14 and 21) provide the capacity for persons to monitor their performance themselves and thus to *learn how to learn*. The second goal of the learning sciences is to generate *models of competences*, which progressively evolve and enable educators to transform learners into efficient problem-solvers (see Chap. 18) through the use of reliable and valid cognitive assessment factors. Examples of this process can be found in the *problem-based learning* and *team-based learning* approaches.

Twenty-First century skills are traits, knowledge, life skills, career skills and habits that are of overarching importance to the success of learners as they go through school and in the world of work. These skills include innovation skills, critical thinking skills, communication skills, problem-solving skills and digital literacy skills, as well as skills for creativity, perseverance, collaboration, literacy, self-direction, and global awareness. Buckle (2022) maintains that twenty-first century skills are important for the following reasons: (1) businesses and educators have continued to cite these skills as being the most important means to success in both the world of work and in academic achievement; (2) since educational institutions prepare students for jobs that may not yet exist, the students have to acquire a nuanced set of skills that prepare them for the future; (3) social media have changed the nature of social interactions and, at the same time, have created emerging challenges that require brand new skills; (4) the Internet has astronomically increased access to knowledge, thus making it necessary for learners to devise ways and means of processing, analysing and evaluating large amounts of information; and (5) there is an increasing realisation that content knowledge of subject matter cannot go very far – the corollary is that learners have to acquire the capacity to apply facts, ideas and principles when solving problems. To assure the acquisition of twenty-first century skills, it is important that educational institutions build staff capacity in order for them to support learners, assess the skills of learners, and provide the teachers with data that enable them to provide support for learners in need of assistance. Part III of the book, comprising seven chapters, anchor the perspectives on the learning sciences and twenty-first century skills.

Summary

In this chapter, I have explained that philosophical foundations of S&T are unique and are concerned with how scientific and technological knowledge are arrived at. The chapter sees curriculum development, hinged on sound philosophical foundations, in terms of the tripod of society, learner and content knowledge to be imparted. In respect of sustainable development and technology in society, the chapter advocates the understanding of the interdependency of the inhabitants of the Earth through the maintenance of a balance between environmental protection and the present and future needs of humanity. The chapter ends with a discussion of the learning sciences and twenty-first century skills.

Recommended Resources

International Institute for Sustainable Development https://www.iisd.org/
International Society of the Learning Sciences https://www.isls.org/
Organisation for Curriculum Development https://www.jstor.org/stable/1169300
Philosophy of Science Association https://www.philsci.org/

References

Buckle, J. (2022). *A comprehensive guide to 21st century skills.* Retrieved from: https://www.pan-oramaed.com/blog/comprehensive-guide-21st-century-skills. Accessed 12.06.22.
Fusco, J. (2020). *What is learning sciences and why does it matter?* Retrieved from: https://digitalpromise.org/2020/03/10/what-is-learning-sciences-and-why-does-it-matter/. Accessed 12.06.22.
Papa, F. J. (2020). Learning sciences theories, principles, and practices comprising a framework for designing a new approach to health professions. *Medical Science Educator, 31,* 241–247. Retrieved from: https://link.springer.com/article/10.1007/s40670-020-01129-2. Accessed 12.06.22.

Ben Akpan, a professor of science education, is the Executive Director of the Science Teachers Association of Nigeria (STAN). He served as President of the International Council of Associations for Science Education (ICASE) for 2011–2013 and currently serves on the Executive Committee of ICASE as the Chair of the World Conferences Standing Committee. Ben's areas of interest include chemistry, science education, environmental education, and support for science teacher associations. He is the editor of *Science Education: A Global Perspective,* published by Springer; co-editor (with Keith S. Taber) of *Science Education: An International Course Companion,* published by Sense Publishers; co-editor with Professor Teresa Kennedy of *Science Education in Theory and Practice* published by Springer; and the editor of *Science Education: Visions of the Future,* published by Next Generation Education. Ben is a member of the Editorial Boards of the *Australian Journal of Science and Technology* (AJST), *Journal of Contemporary Educational Research* (JCER), *Action Research and Innovation in Science Education* (ARISE) Journal, and APEduC *Journal on Research and Practices in Science Education, Mathematics, and Technology –* an electronic scientific-didactic publication of the Portuguese Association of Science Education. He is the recipient of many commendations, prizes, and awards.

Part I
Philosophical Foundations and Curriculum Development

Chapter 2
Nature of Science and Nature of Technology

Steven S. Sexton

Abstract Nature of Science (NOS) and Nature of Technology (NOT) are critical components of science and technology. NOS has a longer and more widely recognised history than NOT, but they work together to support students' learning of how to make informed decisions. This learning occurs when their teachers plan, prepare, and deliver learning opportunities that are based on meaningful learning. Issues and concerns in education arise in NOS and NOT, as there is no checklist of what teachers need to do in order for students to learn. What is agreed in most literature is that students will not learn either NOS or NOT by just doing science or technology. What teachers do matters, and how they do it matters even more as, whether teachers plan it or not, they will be teaching NOS and NOT. Therefore, NOS and NOT are best understood by students if they are explicitly addressed within the context of students' learning of both the concepts and the practices of science and technology. As both NOS and NOT are about what science and technology are, how they work, and how each impact on people and society, both need to be included across all years of learning to allow students the multiple opportunities to develop their understanding. As a result, NOS and NOT should be the unifying component of students' learning, which leads them progressively from simple to increasingly complex knowledge of the world around them, and to how they are able to interact and impact on this world through informed decision-making.

Keywords Nature of science · NOS · Nature of technology · NOT · Education

S. S. Sexton (✉)
College of Education, University of Otago, Dunedin, New Zealand
e-mail: steven.sexton@otago.ac.nz

© The Author(s), under exclusive license to Springer Nature
Switzerland AG 2023
B. Akpan et al. (eds.), *Contemporary Issues in Science and Technology Education*, Contemporary Trends and Issues in Science Education 56,
https://doi.org/10.1007/978-3-031-24259-5_2

Introduction

As the title of this chapter indicates, there is no 'The' nature of science (NOS) just as there is no 'The' nature of technology (NOT). Both NOS and NOT address issues of what each is, how each works, and how society impacts on each. NOS and NOT are different but related. NOS is arguably more widely recognised than NOT, but both share a lack of understanding about effective teaching and learning (Clough et al., 2013).

I am a primary teacher who now works in initial teacher education (ITE). Over the years, I have collected a range of quotes and sayings that have meaning for me and illustrate how I see teaching and learning. One of these is by George S. Patten: *'Say what you mean and mean what you say'*. The issue arises when people use the same word but with different meanings. The final sentence of this chapter's opening paragraph above raises such an issue: '[NOS and NOT] both share a lack of *understanding* about effective teaching and learning'. This phrase assumes that both the readers of Clough et al. (2013) and the authors share the same meaning of the word 'understanding', which includes the grammatical structure of it as the object of a preposition. The word 'understanding' has the infinite form of the verb 'to understand', which has two predominant meanings: one is to be thoroughly familiar with, and the second is to apprehend clearly its character, nature or subtleties. I would argue that Clough et al. (2013) intend for us, as the readers, to apply the second meaning. But we, as readers, also need to know the grammatical structure that places 'understanding' as the object of a preposition. In English grammar, the object of a preposition must be a noun, while 'to understand' is a verb; to be a noun, the authors use the gerund form 'understanding', i.e., the noun form of the verb. Therefore, combining the second definition with 'understanding' as a gerund indicates that many teachers not only do not grasp the character, nature, or subtleties of either NOS and NOT, but also how they are both the concept (i.e., noun) and the action (i.e., verb).

This chapter will provide a brief synthesis of what both NOS and NOT are, how they work, and how society influences each of them. I will draw heavily from two substantial texts that brought together key ideas, authors, and strategies to support effective teaching and learning in both science and technology. McComas edited a book in 2020 that focused on NOS, while Clough *et al* edited a text in 2013 focusing on NOT. As both of these texts bring together a range of authors, for simplicity in this chapter I will make reference to the page in each text where quotes can be located or note the chapter that readers should go to for more on those ideas synthesised here. After exploring what NOS and NOT are and mean, I will discuss their relationship with education to include the importance and significance of indigenous knowledge. After relationships, I will address some of the issues and concerns that have arisen about NOS and NOT in education. The next section will then place both of these in terms of OECD's *Future of Education and Skills 2030* project. The chapter concludes with some final thoughts, recommended sources, summary, and references used.

Nature of Science

'Nature of science (NOS) is not a description of how the natural world works (that's science itself), but rather a description of how the scientific enterprise works' (McComas, 2020, p. 5). McComas and Clough (McComas, 2020, chapter 1) explain that NOS is about what science is, how science works, how science influences both the scientist and society. As such, NOS is fundamental to science education. McComas (McComas, 2020, chapter 2), in his discussions about what should be included in NOS in schools, notes that the learning goals for NOS are not designed for the student but for their teachers, or those who developed the curriculum, or those assessing students' learning. More importantly, McComas argues that we want our students to be able to make informed decisions by knowing NOS. Again, here I argue that we need to make sure that we are all on the same page with what it means to make *informed* decisions. McComas is using an adjective (a descriptive word) to modify the word 'decision'. 'Informed' means based on knowledge, or founded on due understanding of a situation. So, an *informed* decision means a decision made after learning about your options and giving the matter careful thought. McComas, I would argue, is making it very explicit that students and therefore their teachers must know NOS. Knowing NOS is not rote memorisation but, as I have argued elsewhere (Sexton, 2021), is instead based on Ausubel's (1968) meaningful learning.

For learning to be meaningful, students build on prior experiences, adding to their understanding of NOS. McComas, Clough, and Nouri (McComas, 2020, chapter 4) highlighted that students' understanding of science moves from simple to complex. This shift from simple to complex is very nuanced and requires teachers to know what they are doing, how they are doing it, why they are doing it this way, and how their students are engaging in this process. Hattie (2012) argued how and why what teachers do is what matters: *'when teachers see learning occurring or not occurring, they intervene in calculated and meaningful ways to alter the direction of learning to attain various shared, specific, and challenging goals'* (p. 15). Unfortunately, McComas *et al* (McComas, 2020, chapter 4) report that many teachers have NOS misconceptions and therefore this impacts their ability to teach NOS and science content effectively, see César Mora's contribution in Chap. 15 on science and technology teaching strategies.

NOS needs to be taught and learnt with science content in meaningful ways. As noted, McComas (2020) includes numerous chapters on how and why the teaching and learning of both NOS and science content should be meaningful. What is not included is a list of what NOS is, or a list of what teachers should include to teach NOS. Rather, the argument is made that, just as there is no one discipline of science, there is no one NOS. I would direct readers to Clough, Herman and Olsen (McComas, 2020, chapter 13), in which three reasons are given for the importance of NOS in science teaching: NOS supports students in considering, understanding, and accepting science ideas; improves attitudes towards science; and enables students to make more informed socio-scientific decisions.

Clough *et al* (McComas, 2020, chapter 13), like almost all the other authors included in McComas (2020), note that while it is generally accepted that students need to know about both NOS and science content, this is not generally occurring in most classrooms. As I have stated previously, it is what teachers do that matters. Also stated previously, NOS is more widely recognised than NOT, yet, in 2020, McComas presented a book of 39 chapters comprising over 700 pages on the rationales and strategies for NOS in science teaching and learning.

Nature of Technology

Clough (Clough et al., 2013, chapter 18) notes that teachers teach nature of technology (NOT) whether they know it or not. The very act of teaching brings into the classroom the teacher's beliefs, values and opinions on the subject matter. It is not the purpose of this chapter for me to debate any country's, school's, or teacher's intended or hidden curriculum. However, as Clough highlighted, it is not *if* teachers teach NOT, but how well this is done. Shume (Clough et al., 2013, chapter 13) reported on how education is moving from computer skills to more integrated, *'critical, cognitive and problem-solving skills with digital technology and communication tools'* (p. 87). This shift in focus for NOT, similar to that for NOS, is on students being able to apply higher order thinking skills rather than a mastery of technical skills. However, teachers also need to understand NOT as they plan for meaningful learning in technology. Teachers will be better able to plan and present technology in their classrooms, as well as prepare for when technology may actually be interfering with students' learning, if they understand NOT, see Louise Lehane's contribution in Chap. 16 on pedagogical content knowledge in science and technology education.

Kruse (Clough et al., 2013, chapter 17) builds on Shume's chapter and identifies five key ideas of which teachers need to be aware in this multi-faceted construct. Firstly, according to Kruse, technology is both the artefact and the creative process and, as such, teachers need to broaden their understanding to include *'practical knowledge, innovation, human activities, and systems of components'* (p. 346). Beside this duality of NOT, teachers also need to understand how NOT advances. NOT is more often the result of continuous progress rather than cataclysmic upheaval. Kruse points out that, in many cases, it is technology adoption that is more critical to the process than actual development. Secondly, an element of NOT that impacts its adoption is the recognition of how value-laden technology is. There will be winners and losers as technology advances; students and teachers need to be able to take into consideration both sides. Thirdly, there is the awareness of technology limitations. While technology may address some problems, it often raises others. More important is the understanding of how and why technology cannot fix all problems. Fourthly, there are what Kruse refers to as 'trade-offs' with technology. When technology advances, it does so at the expense of other artefacts or processes that may be lost as a result of this advance. Finally, technology impacts on culture.

This impact can be felt not only in the wider culture, but also in the classroom of the student. Kruse goes on to argue that, if teachers are going to make informed decisions about technology, they must take into consideration how technology impacts on thinking and what we do. He then spends the next twenty pages providing evidence and examples that support Hattie's (2012) claim that it is not just that teachers matter but, more importantly, what teachers do that matters.

The Relationship Between NOS and NOT in Education

For both science and technology, it is important that teachers and students understand that there is more than just the content of these two subject areas necessary to education (Clough et al., 2013, chapter 4). In addition to the content of each, there are what Tala (Clough et al., 2013, chapter 4) refers to as *'the broader issues related to the production and justification of scientific knowledge, as well as understanding the impact of science in society, and vice versa'* (p. 51). She then notes that there is a parallel NOT to go along with NOS. More importantly, she argues that an understanding of the interaction between NOT and NOS would support the understanding of each in education. As noted, McComas (2020) does not include a list of what NOS is, but rather strategies to include this in meaningful teaching and learning. Clough et al. (2013) notes that NOT is less well known and Kruse does list some key ideas. Here in New Zealand, we have a government-funded resource that supports the teaching and learning of science across the school years, known as the Science Learning Hub (SLH). The SLH (2021) does provide guidance as to what are key themes of NOS, which map quite well against Kruse's (Clough et al., 2013, chapter 17) NOT (see Table 2.1).

As noted in Table 2.1, NOS includes features of knowledge, values, beliefs, and assumptions concerning both ideas and content areas of science. These are similar, but not identical, to the themes of NOT, as NOT involves design processes with an emphasis on the relationship of technology and society (Clough et al., 2013, chapter 4). Tala (Clough et al., 2013, chapter 4) provides a summary discussion (see pages 55–58) as to how NOS and NOT themes have been derived, noting an extensive

Table 2.1 Comparison of NOS and NOT themes

NOS		NOT
Tentative		Artefact and creativity
Creativity		Adoption is value-laden
Observations or inferences		Limitations
Subjectivity		Trade-offs
Function and relationship between theories and data		Impacts culture
Social and cultural		
Empirically-based		

range of literature. What they both have in common is in terms of education: it is not the goal for students to just memorise ideas, but for them to experience meaningful learning. The difficulty arises when, as Tala notes, *'What one understands a particular NOS or NOT theme to mean depends upon the epistemological views (s)he has adopted or the scientific context under consideration'* (pp. 57–58). This difficulty is compounded, as Tala further notes: *'Any particular definition of a NOS or NOT theme is as tentative, if not more tentative, and context-dependent than scientific and technological knowledge themselves'* (p. 58).

As previously noted, I am a primary teacher who now works in ITE in New Zealand. New Zealand's Māori population (the indigenous people of Aotearoa | New Zealand) have seen the strengthening of *Te Tiriti o Waitangi* (The Treaty of Waitangi – the founding document of Aotearoa | New Zealand that established a partnership between Māori and the Crown). Currently, this partnership is not on an equal footing; however, Māori organisations and groups have made some inroads into equality. One such inroad is that now every teacher in New Zealand must demonstrate a commitment to and personal development in *Te Tiriti o Waitangi* as a compulsory component of teacher registration and registration renewal (Education Council, 2017). This repositioning of *te reo Māori, Māori tikanga me ngā kawa* (indigenous language, customs, and ways of being) in authentic equal partnership is also occurring with other indigenous populations (Sexton, 2019). Both NOS and NOT key themes include culture, which must also recognise the influence and importance of indigenous knowledge and ways of being. It is deliberate on my part that I include indigenous knowledge as part of NOS and NOT in this section about the relationship of NOS and NOT in education, rather than in the next section, which focuses on issues and concerns, see Robby Zidny, Jesper Sjöström and Ingo Eilks' contribution in Chap. 12 on indigenous knowledge and science and technology education.

In 2019, I contributed a chapter to *Science Education: Visions for the Future* (Sexton, 2019). This text wanted authors to look at what science education could or should look like in 2070. In far too many countries, education is not meeting the needs of everyone. Many countries like New Zealand have education systems based on neo-liberalistic policies, where education is a commodity that does not serve everyone with equity. For most indigenous populations, this inequity is their normal. Their traditional ways of being and knowing have *'been ignored, belittled, or commodified into western thought'* (Sexton, 2019, p. 447). However, there are many indigenous populations working to rectify this situation, but it will take time. I noted that, by 2070, *'western and indigenous knowledge will be meaningfully, usefully and respectfully partnered'* (p. 447). What this means is that indigenous knowledge in the classroom is not a problem to be solved, but a resource that should be used. In 2019, I wrote and, in 2023, as I write this, I still believe that:

> As a teacher of science, I do not want a curriculum that promotes science only to scientists. I want a curriculum that increases scientific literacy, contributes to the school community's wellbeing, acknowledges the importance of nature, and makes explicit the interconnections of people, places and things (p. 457).

As a primary teacher, what I would now add to this is that there is more to the curriculum than science. As a primary teacher, I teach the whole curriculum. What I want for my students in science, I would also argue is needed in every other curriculum area. However, specifically, for this chapter we need to include both science and technology.

Issues and Concerns in NOS and NOT in Education

As teachers, we need to promote a more authentic view of the relationship between NOS and NOT. For example, in New Zealand our curriculum document notes how science is both the idea and the action (Ministry of Education, 2007). In 2017, the technology learning area was revised to allow inclusion of digital technology (TKI, 2018). In this revision, our Ministry of Education made explicit that technology as a curriculum area is not only the thought and action, but also includes NOT. However, a less informed reader would probably understand this revision to mean that there is a distinct separation of technology from all other curriculum areas. Technology, like science, mathematics, English, social sciences, health/physical education, the arts, and learning languages, is its own curriculum area in New Zealand, with its own unique characteristics. However, there is a strong relationship between science and technology. Both science and technology impact each other. As one develops and advances, it promotes the development and advancement of the other. In many circumstances, it may be difficult and undesirable to even try to distinguish what is the science and what is the technology.

Kruse (Clough et al., 2013, chapter 17) highlighted that the mistake that New Zealand made in 2017 was not uncommon with regard to technology, NOT and education. Our curriculum also placed an emphasis on digital technology. In fact, our curriculum revision positioned digital technology not only as two-fifths of this revised subject area, but also as the first two areas of technology (TKI, 2018). Fox-Turnbull et al. (2021) published a guide for New Zealand teachers to show how to include technology in education. In this, they support what Kruse noted in 2013 that, for most people, digital technology is the main focus of technology. As a result, Fox-Turnbull *et al* want teachers from across the school years to know how they could include meaningful learning through technology, not just digital technology.

Angle (McComas, 2020, chapter 37) noted what Kruse (Clough et al., 2013, chapter 17) highlighted in that students will not just develop an understanding of NOS or NOT. For many students, this will only happen with targeted and planned experiences. These experiences need to be meaningful learning, i.e., connecting new material to prior knowledge, not rote learning, progressive differentiation, integrated reconciliations, anchoring and practice (Ausubel, 1968; Sexton, 2021). Both Angle and Kruse noted the need for NOS and NOT to connect to students' worlds. Similarly, teacher education must also prepare student teachers to be able to be those teachers who can make necessary calculated decisions to alter the direction of learning (Hattie, 2012). Therefore, neither NOS nor NOT is an add-on, but instead an essential component that both students and teachers use when making informed decisions.

Clough (Clough et al., 2013, chapter 18) was referring to NOT when he made the following statement; however, I would argue that it applies also to NOS:

> Deep and meaningful learning demands assiduous mental engagement. Learners must do more than simply attend to information; they must also overtly connect and compare that information to their prior knowledge. However, as previously noted, even when that kind of mental engagement occurs, learners often interpret and sometimes modify information so that it conforms to what they already think. Conceptual learning often demands not simply adding new information to what learners already think, but altering the way they think about their prior experiences and ideas (p. 380).

Clough then goes on to note that the above is not easy. What we as teachers do matters, how we as teachers engage our students in meaningful learning is why we are professionals and important to the learning process. For most students, if we are not there, learning will not happen. This complexity is further convoluted with the way that most students' learning is assessed. For example, both Programme for International Student Assessment (PISA) and Trends in International Mathematics and Science Study (TIMSS) use pencil-on-paper tests. This form of assessment does not support students in their developing of science or technology ideas or how to understand scientific or technological practices, let alone how practice leads to ideas, see Bulent Çavaş, Pinar Çavaş, and Sengul Anagum's contribution in Chap. 6 on assessment and evaluation in science and technology education.

OECD Knowledge and Skills 2030

In 2015, the OECD launched the *Future of Education and Skills 2030* project. This began with the 'Learning Compass 2030' in 2015, and the 'Teaching Framework 2030' in 2019 (OECD, n.d.-a). Like NOS and NOT, OECD makes explicit that knowledge includes both the idea and the action. The 'Learning Compass 2030' notes that theoretical concepts go hand-in-hand with the understanding gained from the meaningful learning experiences that students undertake. The OECD recognises four different types of knowledge: disciplinary, interdisciplinary, epistemic and procedural. However, these types of knowledge are interconnected and interrelated with skills, attitudes and values, all of which need to be developed interdependently. It is through this interconnected and interrelatedness of knowledge, skills, attitudes and values that students are provided with the opportunities to understand, interpret and apply both knowledge and skills in various situations as they work towards making choices and judgements, as well as displaying behaviours and actions supporting wellbeing of not only themselves as individuals but also societal and environmental wellbeing.

The OECD's *Knowledge and Skills 2030* promotes interrelatedness over learning as a set of discrete units of learning. The OECD goes further to note that: '*Education systems around the world have been moving from defining subjects and required curriculum knowledge as collections of facts, towards understanding disciplines as inter-related systems'* (OECD, n.d.-b, para. 3). I have long held the position that the only bad question that a student can ask is *'why are we doing this?'* While some student questions may not be appropriate to the topic, or are an attempt to sidetrack the learning, there is a reason behind every student's question. 'Why are we doing this?' means that the teacher has made a mistake in setting up the learning or presenting the learning opportunity. Students need to know what they are doing, why they are doing it and, when ready, an understanding of where this learning will take them. Students need to know that they are going to school for a reason and that the teacher knows what they are doing, how they are going to do it, and why they are doing it.

The issue arises when curriculum frameworks do not place significance on NOS and NOT. Without NOS and NOT, how will students learn to understand that they need to be informed in order to make decisions? Olson (Clough et al., 2013, chapter 13), commenting on the American 'Framework for Science Education', argues that it ignores or downplays the significance of philosophy of science and focuses on practices. As already noted, science and technology are both the concept and the practice, as well as both NOS and NOT involving assumptions, biases, values and beliefs. Olson goes on to argue *'that students deserve to know, for example, that observations are theory-laden, that scientific knowledge has a tentative yet durable character, and that theories are not tentative guesses nor do they "grow up" into laws'* (p. 239). She has similar arguments for how NOT is inappropriately presented, *'given [that] the very real consequences of modern technologies that impact all of us and future generations, denying students access to the nature of technology, including its assumptions, limitations, and consequences, is nothing short of educational malpractice'* (p. 242).

Final Thoughts

Understanding both NOS and NOT is important for both teachers and students. It is through NOS and NOT that teachers plan, prepare, and present meaningful learning, as well as how they assess both science and technology in education. As noted by the OECD, knowledge, skills, attitudes, and values are interconnected and inter-related. This means that students need exposure to both the concept and the practice to be able to make informed decisions. Neither NOS nor NOT can be a nice-to-have addition only if time allows. Both are central and necessary to the decision-making process. Students need to learn how they are not only users but should also be questioners and producers of science and technology. Just as what teachers do matters, what students learn that they can do will matter. Therefore, initial teacher education programmes have a responsibility to ensure that future teachers not only understand the importance and value of NOS and NOT, but also are able to apply this understanding, so that they are able to make meaningful decisions about science and technology in education.

Summary

This chapter has addressed several issues and ideas. Nature of Science (NOS) is about what science is, how science works, and how science influences both people and culture; while Nature of Technology (NOT) is about what technology is, how technology works, and how technology influences both people and culture. This chapter makes it explicit that NOS is not the same as NOT, but both are related as NOS has key themes related to NOT key themes. Importantly, it is not about *if* teachers teach NOS or NOT in their classrooms, it is about how well they teach NOS and NOT as it is through NOS and NOT that students learn to make informed decisions.

Recommended Resources

Abd-El-Khalick, F., & Lederman, N. G. (2000). Improving science teachers' conceptions of nature of science: A critical review of the literature. *International Journal of Science Education, 22*(7), 665–701. https://doi.org/10.1080/09500690050044044

Bybee, R. W. (2013). *Translating the NGSS for classroom instruction*. NSTA Press.

Clough, M. P. (2011). Teaching and assessing the nature of science: How to effectively incorporate the nature of science in your classroom. *The Science Teacher, 78*(6), 56–60.

Compton, V. J., & Compton, A. D. (2013). Teaching the nature of technology: Determining and supporting student learning of the philosophy of technology. *Journal of Technology and Design Education, 23*, 229–256.

McComas, W. F., Lee, C. K., & Sweeney, S. (2009). *The comprehensiveness and completeness of nature of science content in the U.S. state science standards.* Paper presented at the National Association for Research in Science Teaching International Conference, Garden Grove, CA.

References

Ausubel, D. P. (1968). *Educational psychology: A cognitive view*. Holt, Rinehart and Winston.

Clough, M. P., Olson, J. K., & Niederhauser, D. S. (Eds.). (2013). *The nature of technology: Implications for learning and teaching*. Sense Publishers.

Education Council. (2017). *Our code, our standards | Ngā tikanga matatika ngā paerewa*. Education Council.

Fox-Turnbull, W., Reinsfield, E., & Forret, A. M. (2021). *Technology education in New Zealand: A guide for teachers*. Routledge.

Hattie, J. (2012). *Visible learning for teachers: Maximizing impact on learning*. Routledge.

McComas, W. F. (Ed.). (2020). *Nature of science in science instruction: Rationales and strategies*. Springer.

Ministry of Education. (2007). *The New Zealand curriculum*. Learning Media.

Organisation for Economic Co-operation and Development (OECD). (n.d.-a). *In brief: Knowledge for 2030*. Retrieved from: https://www.oecd.org/education/2030-project/teaching-and-learning/learning/knowledge/in_brief_Knowledge.pdf

Organisation for Economic Co-operation and Development (OECD). (n.d.-b). *OECD future of education and skills* 2030. Retrieved from:https://www.oecd.org/education/2030-project/about/

Science Learning Hub (SLH). (2021). *Describing the nature of science*. Retrieved from: https://www.sciencelearn.org.nz/resources/412-describing-the-nature-of-science

Sexton, S. S. (2019). Indigenous knowledge. In B. Akpan (Ed.), *Science education: Visions of the future* (pp. 447–462). Next Generation Education.

Sexton, S. S. (2021). Meaning learning – David P. Ausubel. In B. Akpan & T. J. Kennedy (Eds.), *Science education in theory and practice* (pp. 163–176). Springer.

Te Kete Ipurangi (TKI). (2018). *Technology*. Retrieved from: https://nzcurriculum.tki.org.nz/The-New-Zealand-Curriculum/Technology

Steven S. Sexton is a senior lecturer at the University of Otago, College of Education. He obtained his PhD from the University of Sydney in 2007. He has been a classroom teacher in Japan, Thailand, Saudi Arabia, Australia, and New Zealand. Currently, he delivers science education papers in both the undergraduate initial teacher education programme and the Master of Teaching and Learning programme. He has been the Editor of the *Science Education International* journal since 2017. His research interest areas are in relevant, useful, and meaningful learning in science education, teacher cognition, and heteronormativity in schools.

Chapter 3
The Theory of Evolution

Ben Akpan

Abstract This chapter explores the meaning of evolution within the background of the recognition of the phenomenon as the single most dominant theme in present-day biology, and provides basic tenets that undergird its occurrence. This is followed by a discussion of the evidence of evolution through fossil record, anatomy and chemical composition, geographic distribution, and genetic changes, as well as the applications of evolution in medicine, agriculture and industry. There is an overview of the timeline of evolution, as well as some creationists' perspectives of Young Earth Creationism (YEC), Old Earth Creationism (OEC), Intelligent Design (ID) and Theistic Evolution (TE), with a warning that, in every case where creationist beliefs come up, teachers should exercise caution and handle this in such a way as to be respectful of students' views, especially where religious sentiments are apparent. The chapter ends with further advice for teachers to unequivocally impart the message that the theory of evolution is supported by an overwhelming body of evidence and is fully accepted by scientists.

Keywords Evolution · Adaptation · Natural selection · Survival of the fittest · Fossil record · Anatomy · Chemical composition · Genetic changes · DNA code · Humans · Young earth creationism · Old earth creationism · Intelligent design · Theistic evolution

Introduction

The process by which nature selects, from the genetic diversity of a population, traits that would make an individual more likely to survive and reproduce in a continuously changing environment is termed *evolution*. So, over several years and several generations, the full diversity of life on Earth is expressed. Evolution is the

B. Akpan (✉)
The STAN Place, Abuja, Nigeria

© The Author(s), under exclusive license to Springer Nature 25
Switzerland AG 2023
B. Akpan et al. (eds.), *Contemporary Issues in Science and Technology Education*, Contemporary Trends and Issues in Science Education 56,
https://doi.org/10.1007/978-3-031-24259-5_3

single most dominant theme in biology today. It is one of the most fundamental organising principles of the biological sciences. Evolution emphasises the *relatedness* of life rather than its differences. Evolution provides a scientific explanation for why there are so many different kinds of organisms on Earth and how all the organisms on our planet are part of an evolutionary lineage. It demonstrates why some organisms that look different are in fact quite related, while other organisms that may look similar are only distantly related. Evolution, indeed, accounts for the appearance of humans on Earth and shows our biological connections with other living things, as well as detailing how various groups of humans are related to each other and how we acquired the traits that we have. Evolution facilitates the development of effective new ways to protect humanity against bacteria and viruses that continue to evolve. It provides a *framework* through which we study and understand life on Earth and it is also a way of bringing together many diverse aspects of the complexity of life. A characteristic of life is the ability of organisms to *adapt* to their environments as they change over time. For instance, all bacterial pathogens have become at least somewhat resistant to antibiotics over the past six decades or so. Thus, over time, organisms may change in their appearance and other visible characteristics, as well as in their genetic structure. Still, over long periods of time, these changes become significantly different from what they were at the start. Since the changes take 10,000 s to millions of years to occur, no one has witnessed the origin of a major new animal or plant group. Even so, scientists do have an increasing amount of fossil data that show the evolution of one species from another, step by step and, currently, with molecular techniques it is possible to observe and measure the rate of evolution in many species.

The theory of evolution was developed by Charles Darwin in the mid-1800s after a lifetime of travel, observation and experimentation. Darwin made detailed notes on the variations in species as well as their relationship to fossil forms. He also looked at breeds of domesticated animals, such as dogs, and noted the variations caused by selective breeding (*human-directed evolution*). So, if humans can do this in thousands of years, nature can just as well do this given millions of years. Basically, the *theory of evolution by natural selection* hinges on the following ideas:

- all living things consist of a unique combination of chemicals organised in unique ways – variations occur in every species and no two individuals of a species are alike;
- species' populations are able to adapt to gradually changing environments – the same species in different parts of the world have different tolerances and slightly different characteristics to survive the local conditions in which they live;
- most of the variations have a genetic basis – variations can be passed on to the offspring;
- each species produces more offspring than will survive into maturity;
- those individuals whose variations best fit their environment will more likely survive and reproduce – there is a *struggle for existence*, with the *survival of the fittest*;

- by a process of natural selection, evolution sorts these numerous variations within a population and *chooses* the most fit combination – as the environment slowly changes and certain variations are selected over thousands of generations, new forms arise; and
- failure to evolve in response to environmental changes can, and often, does lead to extinction.

Like other foundational theories (for example heliocentric theory, cell theory and the theory of plate tectonics), the theory of evolution is supported by numerous observations and confirming experiments, so much so that scientists are confident that the basic components of the theory will not be overturned by new evidence. However, like all scientific theories, the theory of evolution is subject to continuing refinement as new areas of science emerge, or as new technologies enable observations that were not possible previously. Indeed, the past and continuing occurrence of evolution is a scientific fact: since the evidence supporting it is so strong, scientists no longer question whether *biological evolution* has occurred and is continuing to occur. Instead, scientists investigate the mechanisms of evolution, how rapidly it can take place, as well as other related questions (NAS, 2008).

Evidence for Evolution

There are four primary sources of the evidence for the occurrence of evolution:

Fossil Record Scientists have examined remains of animals and plants that have been found in deposits of sedimentary rocks and have obtained records of past changes through vast periods of time that are impossible to doubt. Such discoveries confirm the fact that there has been a tremendously large variety of living things. Some of the discoveries show species now extinct but that were transitional between some groups of organisms. This shows that species of organisms are indeed not fixed but can, and do, evolve into other species over time. In fact, what have appeared as gaps in the fossil record are actually due to data collection that as yet is incomplete. Scientists are, thus, gradually filling in the *missing links* between transitional fossil specimens as more discoveries are made. O'Neil (2013) reports that one of the first of these gaps that have been filled involved small bipedal dinosaurs and birds. This happened barely two years after Charles Darwin had published *On the Origin of Species*. In this case, a 150–145-million-year-old fossil of *Archaeopteryx* was discovered in Germany. The species had jaws with teeth, as well as a long bony tail similar to dinosaurs, broad wings and feathers similar to those of birds and, interestingly, skeletal features of both dinosaurs and birds. The discovery confirmed that, over time, reptiles evolved into birds. According to O'Neil, after this discovery there have been several other evolutionary gaps that have been filled in the fossil record, the most outstanding one from the human perspective being that between apes and our own species. He reports that, since the 1920s, hundreds of dated intermediate fossils have been found in Africa that were in fact transitional species leading from apes to humans over the last 6–7 million years.

Similarities in Anatomy and Chemical Composition All living things are very similar in their chemical compositions, as well as in their anatomical structures. Scientists have discovered that all living things:

- begin as single cells that, through division processes, reproduce themselves;
- eventually grow old and die;
- share the unique ability to create about 99% of complex molecules of proteins, carbohydrates, fats and other molecules from just 6 (carbon, hydrogen, nitrogen, oxygen, phosphorus and sulphur) of the 118 elements;
- obtain their unique characteristics from their parents, inheriting particular combinations of genes that are actually segments of DNA, which contain coded formulas for creating proteins by linking together particular amino acids in specific order; and
- show evidence of fundamental molecular unity of life, in spite of the enormous diversity of life, in that the basic language of the DNA code is the same for all living things.

Furthermore, many living things show similarities as they derive energy for growth, reproduction and repair directly from sunlight through photosynthesis, or indirectly through the consumption of green plants as well as organisms that feed on plants. Indeed, many groups of species share similar body structures. For example, the arms of humans, the forelegs of dogs as well as cats, the wings of birds, and the flat broad limbs of whales/seals all have the same types of bones – ulna, radius and humerus – as they retained these traits from their shared common ancient vertebrate ancestor.

Geographic Distribution of Related Species Scientists have discovered that major isolated land areas and islands often evolved their own distinct plant and animal species. For instance, before the arrival of humans in Australia about 40,000 years ago, there were none of the more advanced *placental* mammals such as dogs, cats, bears and horses, although there were more than 100 species of kangaroo, koala and other marsupials. In the more isolated islands such as New Zealand and Hawaii, land mammals were entirely absent. Yet, these places had many plant, insect and bird species that were found nowhere else in the world, indicating that the life forms in these areas evolved in isolation from the rest of the world (*ibid*).

Genetic Changes Widespread deaths do occur among species due to environmental changes that most members of the species cannot endure. However, because natural populations do have genetic diversity, not all individuals perish. Individuals with characteristics that allow them to survive adverse environmental conditions will survive and reproduce and pass on their traits to the next generation, thereby ensuring that evolution has occurred. Similarly, the phenomenon of bacterial evolution in the human body is the cause of antibiotic resistance – when an antibiotic medicine is not able to completely cure a bacterial infection. Selective breeding of new varieties of animals and plants also occurs due to environmental changes, as individuals lacking the desirable characteristics are not allowed to breed, with the resulting generations more commonly having the desired traits. Species such as insects and microorganisms that mature and reproduce large numbers in a relatively short period of time have great potential for fast evolutionary changes. This has resulted, for example, in humans' inability to combat the menace of insects as the pesticides used against them become ineffective.

Timeline of Evolution

Here follows the approximate timeline for the evolution of life:

3.8 billion years ago: This is currently the best estimate for the beginning of life on Earth. It is thought that the first life might have developed in undersea alkaline vents based on ribonucleic acid (RNA) instead of deoxyribonucleic acid (DNA).

3.5 billion years ago: Single-celled organisms form.

3.4 billion years ago: Rock formations in Western Australia appear.

3 billion years ago: Viruses emerge.

2.4 billion years ago: *The Great Oxidation Event*, (also called the Oxygen Catastrophe, or the Oxygen Crisis) occurs as cyanobacteria living in the oceans start producing oxygen through photosynthesis. As oxygen builds up in the atmosphere, anaerobic bacteria are killed leading to the Earth's first mass extinction.

2.3 billion years ago: The first *snowball Earth* occurs when the Earth freezes over as a result of volcanic activity. When the ice subsequently melts, more oxygen is released into the atmosphere.

2.15 billion years ago: Evidence of cyanobacteria and photosynthesis emerges.

2 billion years ago: Cells with organelles (eukaryotic cells) come into being.

1.5 billion years ago: The eukaryotes divide into three groups: the ancestors of modern plants, fungi, and animals split into separate lineages, and evolve separately.

900 million years ago: Multicellular life occurs.

800 million years ago: Multicellular animals divide into *Sponges* and *Eumetazoa*. Later, thin plate-like creatures about 1 millimetre across, known as *Placozoa*, break away from the rest of the *Eumetazoa*. *Placozoa* are considered to be the last common ancestor of all animals.

770 million years ago: Earth freezes again.

730 million years ago: Ctenophores split from the other multicellular animals.

680 million years ago: Jellyfish and their relatives break away from the other animals.

630 million years ago: Some animals evolve *bilateral symmetry* – they have a defined top and bottom as well as front and back.

590 million years ago: The Bilateria (animals with bilateral symmetry) split into:

- deuterostomes: these include all vertebrates and ambulacraria; and
- protostomes: comprising all the arthropods (insects, spiders, crabs, shrimp, etc.), worms, and the microscopic rotifers.

540 million years ago: Sea squirts form.

530 million years ago: True vertebrates emerge.

500 million years ago: Animals explore the land.

465 million years ago: Plants grow on land.

460 million years ago: Fish split into two groups – bony fish and cartilaginous fish.

400 million years ago: Insects emerge.

397 million years: Four-legged animals (tetrapods) emerge on land and give rise to amphibians, reptiles, birds and mammals.

375 million years ago: Tiktaalik, an intermediate between fish and tetrapods, emerges.

250 million years ago: Greatest mass extinction in Earth's history occurs. Sauropsids – mostly in the form of dinosaurs, the ancestors of mammals – survive as small, nocturnal creatures.

180 million years ago: The monotremes, mammals that lay eggs rather than giving birth to live young, break apart from the others.

150 million years ago: Archaeopteryx (*first bird*) appears in Europe.

140 million years: Placental mammals appear.

130 million years ago: Flowering plants appear.

100 million years ago: Largest land animal in Earth's history, *Argentinosaurus*, lives around this time.

70 million years ago: Grasses evolve.

63 million years ago: Primates split into two groups:

- haplorrhines – these develop into monkeys, apes and humans; and
- strepsirrhines – these eventually become the modern lemurs and aye-ayes.

50 million years ago: Whales appear.

40 million years ago: New World monkeys appear.

25 million years ago: Apes split from the Old-World monkeys.

14 million years ago: Orangutans emerge from other great apes, spreading across southern Asia, leaving their cousins in Africa.

6 million years ago: Humans diverge from their closest relatives, the chimpanzees and bonobos. Shortly afterwards, hominins begin walking on two legs.

Applications of Evolution

Knowledge of evolutionary trends has been used in many areas of human endeavour, including the following areas:

Medicine An understanding of evolution has been applied widely in the medical field. For example, the identification of the *severe acute respiratory syndrome*

(SARS) in 2002 was facilitated by using a technique, the *DNA micro-array*, which is based on knowledge of evolutionary trends. The technique identified the virus as a previously unknown member of a particular family of viruses. The genetic material in the SARS virus was similar to that of other viruses because it had evolved from the same ancestor. Following the identification, blood tests were carried out to identify people with the disease. Further efforts resulted in the identification of appropriate medicines for treatment of infected persons and the production of vaccines to prevent future infections. Knowledge of evolutionary pathways of viruses will be useful in future as these pathogens evolve into more resistant forms.

Agriculture   Scientists have applied an understanding of evolution to find out the relationships existing among plants and to identify the traits that can be employed in the improvement of crops. They have therefore been identifying the genes in the DNA of plants that are responsible for their advantageous traits so that these traits can be incorporated into other crops. Indeed, processes of evolutionary change have been used to transform many wild plants and animals into crops and domesticated animals. Farmers do save seeds from plants with particularly favourable traits. They later plant those seeds in the next growing season. This process of artificial selection creates a variety of crops with characteristics particularly suited for agriculture. According to NAS (2008), farmers have modified wild wheat so that seeds remain on the plant when ripe and so can be separated with ease from their hulls.

Industry   Natural selection principles have found wide applications in industries. Chemists, for instance, have applied these principles in the development of new molecules with specific functions. They start by creating variants of an existing molecule. The variants are then tested for the desired function. New variants are generated from the variants that do the best job. By continually repeating this selection process, chemists obtain molecules that have a greatly enhanced ability to perform a particular task. For example, new enzymes that can convert cornstalks and other agricultural waste into ethanol with improved efficiency have been created through the application of this technique.

Creationist Perspectives

Creationism is the belief that life originated through a process governed by a supernatural entity. In contrast, the theory of evolution holds that humans and other species are products of natural selection and random mutation that gradually, over long periods of time, produce life forms that are more complex from simple life forms. Henry Morris is generally regarded as the father of modern creationism. Dr. Morris, a hydraulic engineer, in liaison with the theologian, John Witcomb, published *The Genesis Flood* in 1961, which has been widely regarded as the handbook of creationism (Rudoren, 2006). There are different perspectives on creationism and it is to these that I now turn:

Young Earth Creationism (YEC)   Young Earth Creationists believe that the Earth, as well as its lifeforms, were created in their present forms by a divine entity approximately 6000 to 10,000 years ago. YEC is thus based on literal interpretations of the Book of Genesis.

Old Earth Creationism (OEC)   Old Earth Creationists maintain that the Earth and its lifeforms were created by a supernatural entity, much like the YEC. However, OEC accepts the scientific evidence for the age of the Earth and of creation in the Book of Genesis as being of unspecified length, thus stretching out to fit the results of scientific studies.

Intelligent Design (ID)   Intelligent Design accepts an old Earth and most science, but maintains that some features of living things as well as the universe are best explained from the perspective of an *intelligent cause* rather than an *undirected process* such as natural selection. So, while science provides two explanations for evolutionary trends (that is, necessity or *natural law* and chance or *variation*), ID adds a third explanation, *design*, which takes it out of the world of science. ID, thus, takes the position that scientific explanations are insufficient in explaining the apparent design in nature. ID does not accept that genetic mutation and natural selection (both totally unguided processes) could have brought about life as we know it. ID, therefore, postulates the intervention of a transcendent intelligent designer as the best explanation. Mulherin (2014) thinks that the argument for ID is more of the *god-of-the-gaps* argument and that, as science develops, the gaps in knowledge shrink and squeeze out the need for a designer. Indeed, the tenets of ID are not in agreement with scientific findings. Scientists have examined the various molecular systems that are claimed to result from design and have found that they could have arisen through natural processes. NAS (2008) weighs in:

> In the case of the bacterial flagellum, there is no single, uniform structure that is found in all flagellar bacteria. There are many types of flagella, some simpler than others, and many species of bacteria do not have flagella to aid in their movement. Thus, other components of bacterial cell membranes are likely the precursors of proteins found in various flagella. This similarity indicates a common evolutionary origin where small changes in the structure and organisation of secretory proteins could serve as the basis for flagellar proteins. Thus, flagellar proteins are not irreducibly complex (pp. 40–41).

Indeed, existing systems may be capable of acquiring new functions such that a particular system having one task in a cell becomes adapted through the process of evolution for a different use. The *Hox genes*, a family of regulatory genes that encode transcription factors and are essential during embryonic development, are an important case demonstrating how evolution finds new uses for existing systems. Molecular biologists have discovered that gene duplication provides an important pathway in which biological systems acquire additional functions. It is, however, important to note, as Poole (2008) opines, that, in the interests of proper science education, the rejection of the *ID argument for design* as a bad argument should not be presented as dismissing the *traditional belief in design* itself. This is because biological evolution anchored on the concept of *chance* and *selection* does not preclude design, especially in consideration of genetic algorithms where experts use computers to mimic the molecular processes involved in sexual reproduction to work out optimum conditions for solving a wide range of problems.

Theistic Evolution (TE)   TE takes the position that, although evolution occurred, a Creator or intelligence was involved in the process. It is the view of those who believe that God is responsible for life, the universe and everything ultimately, but who also accept the findings of science. While they accept that evolution is the best explanation of the data, they nonetheless do not accept the naturalistic philosophy that embodies scientific explanation.

Concluding Remarks

Biological evolution provides the key to understanding the principles governing the origin and extinction of living things. It allows scientists to determine how and why organisms have become the way we find them, as well as the processes currently acting to modify their present state. Evolutionary biology has the capacity to contribute to humanity's awareness of the consequences of environmental disturbances such as deforestation, application of pesticides and global warming. Biological evolution is therefore one of the most important ideas of modern science, as it provides the basis of the modern biological sciences with applications as well in many other scientific and engineering fields. Since evolution has the potential to serve as an important foundation of some key science disciplines, it is important that students are assisted to learn about and understand the evidence, mechanisms and implications that undergird it. Even so, it is sometimes the case that, when teaching evolution, situations arise indicating some doubts among the students based on cultural or religious backgrounds. Reiss (2008) advises that if questions or issues about creationism and intelligent design arise in the course of a lesson, such opportunities could be utilised to illustrate a number of aspects of how science works, to wit: need for evidence to test ideas and develop scientific theories; that there are some questions that science cannot currently answer; that scientific knowledge and ideas change over time; and that the scientific community plays a leading role in validating these changes. In every case where creationist beliefs come up, teachers should exercise caution and handle the situation in such a way as to be respectful of students' views, especially where religious sentiments are apparent. However, the teacher should unequivocally impart the message that the theory of evolution is supported by an overwhelming body of evidence and is fully accepted by scientists.

Summary

In this chapter, I have discussed the meaning and tenets of evolution as well as evidence, timeline, and applications of evolution. Evolution is the single most dominant theme in biology today. It is one of the most fundamental organising principles of the biological sciences. It emphasises the relatedness of life rather than its differences. Evolution provides a scientific explanation for why there are so many different kinds of organisms on Earth and how all the organisms on our planet are part of an evolutionary lineage. Other topics discussed in the chapter include young Earth creationism, old Earth creationism, intelligent design, and theistic evolution.

Recommended Resources

European society for evolutionary biology https://www.scientia.global/
    the-european-society-for-evolutionary-biology/
Society for integrative and comparative biology https://sicb.org/
Society for the study of evolution https://www.evolutionsociety.org/

References

Mulherin, C. (2014). *Categories of creationists...and their views on science.* Retrieved from:
    https://theconversation.com/categories-of-creationists-and-their-views-on-science-27123.
    Accessed 28.05.22.
NAS (National Academy of Sciences). (2008). *Science, evolution, and creationism.* The National
    Academies Press.
O'Neil, D. (2013). *Evidence of evolution.* Retrieved from: https://www2.palomar.edu/anthro/
    evolve/evolve_3.htm. Accessed 12.10.21.
Poole, M. (2008). Creationism, intelligent design and science education. *School Science Review,*
    90(330), 123–129.
Reiss, M. J. (2008). Teaching evolution in a creationist environment: An approach based on world-
    views, not misconceptions. *School Science Review, 90*(331), 49–56.
Rudoren, J. (2006). *Henry M. Mories, 87, a theorist of creationism, dies.* Retrieved from: https://
    www.nytimes.com/2006/03/04/us/henry-m-morris-87-a-theorist-of-creationism-dies.html.
    Accessed 28.5.22.

Ben Akpan, a professor of science education, is the Executive Director of the Science Teachers
Association of Nigeria (STAN). He served as President of the International Council of Associations
for Science Education (ICASE) for 2011–2013 and currently serves on the Executive Committee
of ICASE as the Chair of the World Conferences Standing Committee. Ben's areas of interest
include chemistry, science education, environmental education, and support for science teacher
associations. He is the editor of *Science Education: A Global Perspective* published by Springer;

co-editor (with Keith S. Taber) of *Science Education: An International Course Companion* published by Sense Publishers; co-editor with Professor Teresa Kennedy of *Science Education in Theory and Practice* published by Springer; and the editor of *Science Education: Visions of the Future* published by Next Generation Education. Ben is a member of the Editorial Boards of the *Australian Journal of Science and Technology* (AJST), *Journal of Contemporary Educational Research* (JCER), *Action Research and Innovation in Science Education* (ARISE) Journal, and *APEduC Journal on Research and Practices in Science Education, Mathematics, and Technology* – an electronic scientific-didactic publication of the Portuguese Association of Science Education. He is the recipient of many commendations, prizes and awards.

Chapter 4
STEM Education as a Meta-discipline

Teresa J. Kennedy and Michael R. L. Odell

Abstract STEM education has evolved into a meta-discipline, an integrated effort that removes the traditional barriers between the subjects of science, technology, engineering and mathematics, and focuses on innovation and the applied process of designing solutions to complex contextual problems. Stakeholders including schools, community organizations and businesses acknowledge the fundamental links between economic prosperity, knowledge-intensive jobs dependent on science and technology, and the importance of building a culture of continued innovation aimed at addressing current societal problems and those of the future. STEM education provides students with the knowledge and skills that they need to be successful in the twenty-first century. This chapter explores STEM as a meta-discipline, discusses the origins and emergence of STEM education, as well as the differences between S.T.E.M and STEM, and the many variations of the acronym. STEM education policy development in countries around the world, definitions of STEM literacy, and pedagogical implementation models are also described.

Keywords Science · Technology · Engineering · Mathematics · STEM · STEM education · Interdisciplinary frameworks · Science pedagogy · PISA · Twenty-first century skills

Introduction

Over the last 25 years, significant funding, research, and development have focused on STEM education on a global scale. What is STEM education? STEM is the acronym for the subject areas of science, technology, engineering and mathematics (STEM). These four STEM umbrella disciplines are deemed by governments,

T. J. Kennedy (✉) · M. R. L. Odell
The University of Texas, Tyler, TX, USA
e-mail: tkennedy@uttyler.edu; modell@uttyer.edu

© The Author(s), under exclusive license to Springer Nature
Switzerland AG 2023
B. Akpan et al. (eds.), *Contemporary Issues in Science and Technology Education*, Contemporary Trends and Issues in Science Education 56,
https://doi.org/10.1007/978-3-031-24259-5_4

business and industry, and educators as essential to the global economy, competitiveness in the workforce, and in education. STEM disciplines "typically include educational activities across all grade levels—from pre-school to post-doctorate—in both formal (e.g., classrooms) and informal (e.g., after-school programs) settings" (Gonzalez & Kuenzi, 2012, p. 2).

Many different teaching approaches have been implemented to improve student learning in STEM disciplines. For example, numerous educators have employed project-, problem-, and/or phenomenon-based learning activities that require knowledge and skill application in specific areas, such as engineering. In addition, extracurricular activities are offered, such as team competitions encouraging student collaboration (e.g., solving a problem or apply coding and engineering principles to design and build robots). In these situations, students are typically afforded time with professionals in STEM fields (e.g., teachers organize guest lectures by STEM experts; STEM professionals provide feedback on class projects and/or serve as science fair judges providing feedback on student innovations; local companies provide students with organized opportunities to job-shadow and work as interns; etc.).

STEM educational opportunities also extend beyond the school classroom environment, occurring at home, in other venues such as the wide genre of museums, and during leisure time in the learners' communities (Kennedy & Tunnicliffe, 2022, p. 12). STEM-focused initiatives are intentionally planned and implemented around the world in a variety of settings by stakeholders such as schools, community organizations, and businesses as a way of building interest and fostering a diverse STEM workforce. These stakeholders acknowledge the fundamental links between economic prosperity, knowledge-intensive jobs dependent on science and technology, and the importance of building a culture of continued innovation aimed at addressing current societal problems and those of the future; since the reality is that there will undoubtedly be jobs that will be necessary in the future that do not exist today.

STEM education, as a meta-discipline, marks an integrated effort to remove the traditional barriers between the content areas of science, technology, engineering and mathematics, and focuses on innovation and the applied process of designing solutions to complex contextual problems. Using current tools and technologies, STEM education challenges students of all ages to innovate and invent, while promoting problem-solving and critical thinking skills that can be applied to their academic as well as everyday lives.

Scientific inquiry generally involves the formulation of a question that can be answered through *investigation*, while engineering design involves the formulation of a problem that can be solved through *design*. STEM education naturally brings these two concepts—investigation and design—together through all four disciplines (Kennedy & Odell, 2014, p. 247). However, STEM as a discipline was not always referred to as 'STEM'. There have been various acronyms, such as SMET, the first U.S. reference created in 2001 by the National Science Foundation (NSF), which referenced the standards that educators should follow, including the U.S. Next Generation Science Standards (NGSS Lead States, 2013) when teaching science, mathematics, engineering, and technology to K–12 students (ages 5–18), along with the inherent skills of analytical thinking, problem-solving and science competencies. NSF rearranged the order of the disciplines to form the acronym STEM later that same year, and many different iterations of the acronym have been created and implemented by countries across the globe. Examples of acronyms utilized will be discussed at length in a later section of this chapter.

Historical and Theoretical Background

One can argue as to the origins and emergence of STEM education. History relates the pioneering efforts of ancient Egyptians and Greeks, along with many others from around the world, who introduced and implemented different aspects of the field of science, establishing the importance of conducting investigations directly involving one's environment through efforts to make sense of their world and gathering evidence to support in-depth understandings. Examples include Ibn al-Haytham (Islamic medieval mathematician and astronomer who paved the way for the modern science of physical optics), Moustafa Mosharafa (the Egyptian theoretical physicist who contributed to the development of Einstein's quantum theory), the physician Hippocrates (known as the father of modern medicine), Aristotle (the inventor of the field of formal logic who identified the various scientific disciplines and further explored the diverse relationships between them), Thales (the infamous mathematician and astronomer who first investigated the basic principles, questioning the originating substances of matter), Empedocles (the philosopher known for a view of matter composed of the four elements of fire, air, water, and Earth), the Turkish astronomer Anaximander (who studied topics related to the fields we now refer to as geography and biology), and of course the study of mathematics by the Pythagoreans.

The concept of engineering has also existed since ancient times, as documented by structures such as the Leshan Giant Buddha and the Great Wall of China, the Chand Baori of India, the underground churches of Lalibela in Ethiopia, the Parthenon in Greece, the Roman Colosseum in Italy, the Aqueduct of Segovia in Spain, The Lost City of Mohenjo Daro in Pakistan, the Egyptian Pyramids and the Teotihuacan Pyramids of Mexico, along with many other UNESCO World Heritage sites such as Machu Picchu in Peru, all renowned as buildings that incorporate sophisticated astronomical alignments.

Science and mathematics have always been an essential part of the curriculum, dating back well into the last two centuries, especially at the university level where subjects such as agricultural science had a direct impact on workforce needs prevalent in society. The case can also be made that the modern STEM curriculum was partially a response to World War II and the need for innovations, medicine, and weapons. STEM overtly emerged during the Cold War with the launch of the Russian satellite, Sputnik. The U.S. government, in response to Sputnik, invested heavily in K–12 science and mathematics education with the goal of building a STEM workforce to support national security and the Space Race.

After the Cold War, STEM emerged as a driving factor of economic competitiveness in the global workforce. As the global economy became more interdependent, STEM education became a focus in most countries worldwide. For example, from 2000–2010, STEM jobs in the United States grew at a rate three times greater than other occupations. Due to the shortage of STEM professionals globally, policies have been instituted to increase the STEM pipeline in education and expand the diversity of the workforce, by creating educational programming targeting underrepresented groups including minorities and women. For more information, see Chap. 10, which discusses gender and equity in science and technology education in the context of education for sustainable development.

STEM education and research are necessary requirements for a nation's development, productivity, competitiveness, and societal wellbeing (Marginson et al., 2013). Evidence that this is the case can be found in government efforts worldwide to improve STEM education at all levels—primary, secondary, and tertiary.

STEM education plays an increasingly important role in countries' economic wellbeing and global competitiveness. The Organisation for Economic Co-operation and Development (OECD) ranks achievement by country in mathematics and science using the Program for International Student Assessment (PISA). PISA measures 15-year-old students' abilities to use their reading, mathematics, and science skills to meet real-life challenges (https://www.oecd.org/pisa/).

PISA test questions measure student ability to draw on knowledge and real-world problem-solving skills and, therefore, researchers have concluded that PISA is an important indicator of whether school systems are effectively preparing their students for success in the global knowledge economy of the twenty-first century. The "OECD conducts research on the 65 countries that make up 90 percent of the world's economies. The OECD Directorate for Education has found that student achievement in math and science are a sound indicator for future economic health. In other words, nations or cities with good schools can expect a healthy economy, whereas a nation or city with suffering schools can expect negative consequences to its economy" (Asia Society, n.d., para. 6).

PISA data, collected every 3 years, are often used for benchmarking aimed at driving reform and school turnaround efforts. Knowledge about a nation's placement on this assessment typically affects STEM education policy and provides examples of exemplary educational models. For example, investigating programming in the top five countries (Estonia, Canada, Finland, Ireland, and Korea, in their respective order of accomplishment) reveals common patterns among the top-performing systems and undoubtedly yields replicable implementation models that countries striving to improve could adopt. The most recent PISA results (2019) are depicted in Fig. 4.1.

Although the OECD member countries and associates postponed the PISA 2021 assessment to 2022 and the PISA 2024 assessment to 2025 due to post-COVID difficulties, their 2022 focus will remain on mathematics, with an additional test of creative thinking. "PISA 2025 will focus on science, and include a new assessment of foreign languages. It will also include the innovative domain of Learning in the Digital World which aims to measure students' ability to engage in self-regulated learning while using digital tools" (OECD, n.d.).

Table I.1 [1/2] **Snapshot of performance in reading, mathematics and science**

Countries/economies with a mean performance/share of **top performers above** the OECD average
Countries/economies with a share of **low achievers below** the OECD average

Countries/economies with a mean performance/share of top performers/share of low achievers **not significantly different** from the OECD average

Countries/economies with a mean performance/share of **top performers below** the OECD average
Countries/economies with a share of **low achievers above** the OECD average

	Mean score in PISA 2018			Long-term trend: Average rate of change in performance, per three-year-period			Short-term change in performance (PISA 2015 to PISA 2018)			Top-performing and low-achieving students	
	Reading	Mathematics	Science	Reading	Mathematics	Science	Reading	Mathematics	Science	Share of top performers in at least one subject (Level 5 or 6)	Share of low achievers in all three subjects (below Level 2)
	Mean	Mean	Mean	Score dif.	Score dif.	Score dif.	Score dif.	Score dif.	Score dif.	%	%
OECD average	487	489	489	0	-1	-2	-3	2	-2	15.7	13.4
Estonia	523	523	530	6	2	0	-4	4	-4	22.5	4.2
Canada	520	512	518	-2	-4	-3	-7	4	-10	24.1	6.4
Finland	520	507	522	-5	-9	-11	-6	-4	-9	21.0	7.0
Ireland	518	500	496	0	0	-3	-3	-4	-6	15.4	7.5
Korea	514	526	519	-3	-4	-3	-3	2	3	26.6	7.5
Poland	512	516	511	5	5	2	6	11	10	21.2	6.7
Sweden	506	502	499	-3	-2	1	6	8	6	19.4	10.5
New Zealand	506	494	508	-4	-7	-6	-4	1	-5	20.2	10.9
United States	505	478	502	0	-1	2	8	9	6	17.1	12.6
United Kingdom	504	502	505	2	1	-2	6	9	-5	19.4	9.0
Japan	504	527	529	1	0	-1	-12	5	-9	23.3	6.4
Australia	503	491	503	-4	-7	-7	0	-3	-7	18.9	11.2
Denmark	501	509	493	1	-1	0	1	2	-9	15.8	8.1
Norway	499	501	490	1	2	1	-14	-1	-8	17.8	11.3
Germany	498	500	503	3	0	-4	-11	-6	-6	19.1	12.8
Slovenia	495	509	507	2	2	-2	-10	-1	-6	17.3	8.0
Belgium	493	508	499	2	-4	-3	-6	1	3	19.4	12.5
France	493	495	493	0	-3	-1	-7	2	-2	15.9	12.5
Portugal	492	492	492	4	6	4	-6	1	-9	15.2	12.6
Czech Republic	490	499	497	0	-4	-4	3	7	4	16.6	10.5
Netherlands	485	519	503	-4	-4	-6	-18	7	-5	21.8	10.8
Austria	484	499	490	-1	-2	-6	0	2	-5	15.7	13.5
Switzerland	484	515	495	-1	-2	-4	-8	6	-10	19.8	10.7
Latvia	479	496	487	2	2	-1	-9	14	-3	11.3	9.2
Italy	476	487	468	0	5	-2	-8	-3	-13	12.1	13.8
Hungary	476	481	481	-1	-3	-7	6	4	4	11.3	15.5
Lithuania	476	481	482	2	-1	-3	3	3	7	11.1	13.9
Iceland	474	495	475	-4	-5	-5	-8	7	2	13.5	13.7
Israel	470	463	462	6	6	3	-9	-7	-4	15.2	22.1
Luxembourg	470	483	477	-1	-2	-2	-11	-2	-6	14.4	17.4
Turkey	466	454	468	2	4	6	37	33	43	6.6	17.1
Slovak Republic	458	486	464	-3	-4	-8	5	11	3	12.8	16.9
Greece	457	451	452	-2	0	-6	-10	2	-3	6.2	19.9
Chile	452	417	444	7	1	1	6	5	-3	3.5	23.5
Mexico	420	409	419	2	3	2	-3	1	3	1.1	35.0
Colombia	412	391	413	7	5	6	-13	1	-2	1.5	39.9
Spain	m	481	483	m	0	-1	m	-4	-10	m	m

Notes: Values that are statistically significant are marked in bold (see Annex A3).
Long-term trends are reported for the longest available period since PISA 2000 for reading, PISA 2003 for mathematics and PISA 2006 for science.
Results based on reading performance are reported as missing for Spain (see Annex A9). The OECD average does not include Spain in these cases.
Countries and economies are ranked in descending order of the mean reading score in PISA 2018.
Source: OECD, PISA 2018 Database, Tables I.B1.10, I.B1.11, I.B1.12, I.B1.26 and I.B1.27.
StatLink ⟐ https://doi.org/10.1787/888934028140   ...

Fig. 4.1   Short-term and long-term trends for each participating PISA country

Government efforts worldwide develop policy-level documents guided by PISA data, as well as data associated with the Trends in International Mathematics and Science Study (TIMMS) gathered by the U.S. National Center for Educational Research, and other country-specific data gathered nationally, to govern their unique school STEM education initiatives. Countries often create policy documents aimed at addressing the 'STEM crisis' with regard to unmet labor market demands for STEM skills and the need to remain competitive in the global economy. Freeman et al. (2019) conducted country comparisons and cited several policy reports aimed at the development of a national STEM workforce, such as Australia's 2015 *National STEM School Education Strategy* and 2018 plan *Australia 2030: Prosperity Through Innovation*; New Zealand's *National Statement on Science Investment 2015–2025*; the UK's *Science & Innovation Investment Framework 2004–2014* and 2017 paper, *Industrial Strategy: Building a Britain for the Future;* and, in the United States, the *Rising Above the Gathering Storm* (2007) and *Revisiting the STEM Workforce* (2015) reports.

Their study also revealed that many countries in Western Europe, such as Germany, France, Ireland, the Netherlands, and Spain, have adopted science policies that typically address K–12 school-based science and mathematics teaching, while countries in East and Southeast Asia that have high performing educational systems tend to focus on national policies and plans emphasizing university science and technology programming, industry-driven research and development, and innovation. In addition, they found that emerging economies and education systems, including Brazil, Argentina, and arguably South Africa, have established national policies focused on quality education and emerging industry development. For example, Brazil's *Education Development Plan 2011–2020* emphasizes school education, teaching quality and teacher career pathways; and Argentina's *National Plan of Science, Technology, and Innovation: Argentina Innovadora 2020* prioritizes research and innovation, general scientific capacity, and development of biotechnology and health.

Economic development, including the ability to invent and develop new products, drives the policies of many countries. For example, South Africa's *National Development Plan 2030 of the National Planning Commission* aims to "redress injustices of the past, facilitate economic growth, and improve education, health, and social protection" (p. 6). Additionally, the government of the UAE has promoted STEM fields through its *Vision 2021, Vision 2030, the fourth industrial revolution, and artificial intelligence strategy* (Al Murshidi, 2019, pp. 327–328). These examples clearly demonstrate the diverse STEM policy objectives that generally reflect national cultural, social, and economic/workforce contexts, as well as the need to build strong foundations for STEM literacy and increase diversity, equity, and inclusion in STEM.

With the goal of bringing all countries together, the United Nations' 2030 Agenda for Sustainable Development established 17 Sustainable Development Goals (SDGs) aimed at tackling global issues, including those related to STEM education (see Chap. 10 for more information). UNESCO's International Bureau of Education subsequently developed a resource to identify and describe the contributory elements of STEM competencies associated with the four core STEM subjects, along with potential approaches to teaching in a competency-based manner in order to integrate the four STEM disciplines and create a connected field of study (UNESCO International Bureau of Education, 2019). They found that education, and particularly STEM education, plays a critical role in achieving the internationally-agreed upon outcomes associated with the SDGs, since STEM education aspires to elaborate and provide innovative solutions to solve global issues, in particular those directly related to SDG 2 (Zero Hunger); SDG 3 (Good Health and Well-Being); SDG 6 (Clean Water and Sanitation); SDG 7 (Affordable and Clean Energy); SDG 9 (Industry, Innovation and Infrastructure); SDG 12 (Responsible Consumption and Production); SDG 13 (Climate Action); SDG 14 (Life Below Water); and SDG 15 (Life on Land). Moreover, SDG 8 (Decent Work and Economic Growth) and SDG 11 (Sustainable Cities and Communities) are heavily dependent on progress that can be made within the fields of STEM. In the context of Industry 4.0, the contribution of STEM to achieve the SDGs is crucial (UNDP, 2019).

S.T.E.M Education Versus STEM Education

There are multiple definitions of STEM education. The most simplistic of these is STEM as individual subjects, represented as S.T.E.M. going forward. In this configuration, S.T.E.M. simply refers to the individual disciplines. Another view is that STEM is a meta-discipline. This will be represented as STEM going forward. STEM as a meta-discipline focuses on the interconnected nature of the STEM disciplines versus individual implementation of the four disciplines. STEM as a meta-discipline views STEM as a connected and potentially integrated field of study. Both national and international policymakers have advocated for a STEM agenda that increasingly focuses on the need for STEM concepts in the context of the workplace. As a result, STEM educators need to prepare students to have greater S.T.E.M. literacy and a better understanding of how these STEM disciplines are interconnected and relate to one another.

The central tenet of STEM education is the use of STEM knowledge to solve real-world problems. This can be achieved by adopting the definition that STEM is a meta-discipline unto itself and is delivered in a manner that makes STEM learning more meaningful and contextual. According to Bybee (2013, p. 5), STEM literacy is defined as:

- Knowledge, attitudes, skills [and values] to identify questions and problems in life situations. Explain the natural and designed world, and draw evidence-based conclusions about STEM-related issues;
- Understanding of the characteristic features of STEM disciplines as forms of human knowledge, inquiry, and design;

- Awareness of how STEM disciplines shape our material, intellectual, and cultural environments; and
- Willingness to engage in STEM-related issues with the ideas of science, technology, engineering, and mathematics as a constructive, concerned, and reflective citizen.

It should be noted that the STEM perspective adopted and the operational definition that is used to guide policy and instruction have direct educational consequences. If one is using the perspective of S.T.E.M. as simply four independent disciplines, it is unlikely that an interdisciplinary pedagogical approach would be utilized and would appear as a traditional approach to delivering content to students at the primary, secondary and tertiary levels. The S.T.E.M. approach has been the traditional implementation model in most schools globally until recently.

Advocates for delivering STEM as an interdisciplinary approach believe in removing the disciplinary silos and bringing the disciplines together to form a more applied science or meta-discipline. The alignment of the four STEM areas was first proposed in the 1990s by the National Science Foundation in the United States. As discussed earlier, this approach was proposed to address the demand for STEM skills and competencies that did not result from the traditional approach of S.T.E.M. According to Zilberman and Ice (2021), "Science, Technology, Engineering and Mathematics (STEM) occupations are projected to grow over two times faster than the total for all occupations in the next decade. The U.S. Bureau of Labor Statistics (BLS) 2019–2029 employment projections show that occupations in the STEM field are expected to grow 8.0 percent by 2029, compared with 3.7 percent for all occupations" (p. 1).

In the United States and globally, there have been concerns since the late 1970s and early 1980s about the decline in enrollments and involvement in STEM fields at schools, universities, and colleges (Aina & Akanbi, 2013; Milner et al., 1987; Sithole et al., 2017). This decline in enrollment and motivation to pursue STEM academically has been attributed in part to poor pedagogy, content that was not meaningful, and a lack of applicability to real-world contexts, which resulted in students not relating the knowledge and skills that they learned in school to the real world and their future career choices (UNESCO, 2017).

The framework of STEM presented in the UNESCO report *Exploring STEM Competencies for the twenty-first Century* (UNESCO International Bureau of Education, 2019) was organized around societal needs and how the STEM disciplines work together to fulfill those needs. This approach makes STEM relevant to students, as it provides them with theoretical foundations that enable them to propose timely and innovative solutions to issues and problems confronted by society and the world. Figure 4.2 shows the relationship between the STEM components and depicts how the four STEM components work together to meet the societal needs.

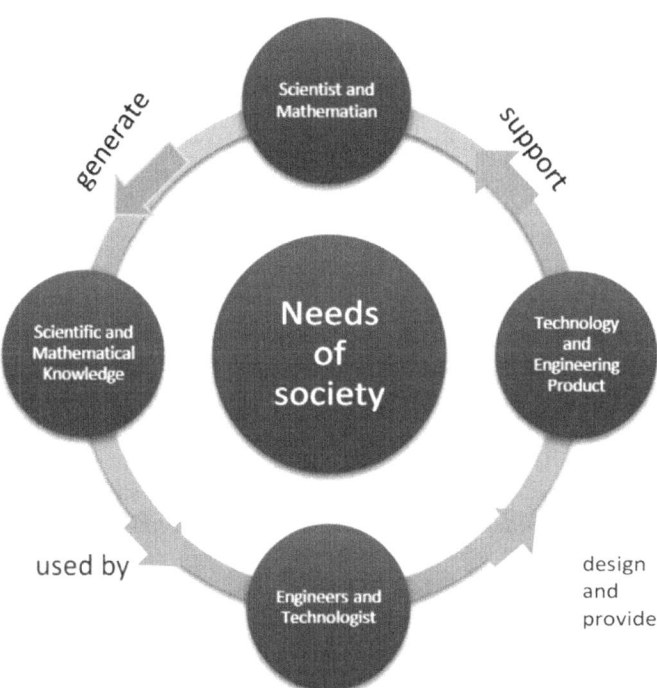

Fig. 4.2 Relationship between the components of STEM (UNESCO International Bureau of Education, 2019, p. 8)

The Evolution of STEM Education

STEM education implementation models continue to evolve. Expanded iterations of the acronym have included additional disciplines as a more holistic approach to education, focusing on individual students' needs and interests. As an example, STEAM education highlights the pedagogical approach that incorporates science, technology, engineering, the Arts and mathematics as access points for student inquiry, creativity, critical thinking and communication. STEAM educators believe that the Arts serve as a mechanism that allows children to incorporate artistic expression to communicate and make better sense of their learning, providing expanded opportunities to reflect, imagine, create, express and represent ideas. They argue that this approach increases student interest and engagement in STEM subjects by integrating creative Arts with various aspects of inquiry-based teaching and learning, including the role of engineering. MacDonald et al. (2019) found that focusing on the spaces between STEM disciplines, the intersections of interdisciplinary study, allows students to discover new insights and gain better understanding of the content areas that they study. According to Holbrook et al. (2020), STEAM education moves beyond disciplines and incorporates a transdisciplinary focus, identifying new knowledge about what is "between, across, and beyond disciplines" (p. 470), further interrelating science education to the relevance and issues of society on local, national, and global scales.

STREAM, an expanded iteration of STEAM education, adds the disciplines of reading and writing. Advocates of STREAM contend that literacy is an essential part of a well-rounded curriculum, requiring critical thinking and creativity (see Chap. 19 for more information on creativity and innovation in science and technology education). Others believe that education should direct itself toward design and design thinking, thereby coining the acronym STEAMD (Henebery, 2020). However, STEAMD has also taken on another definition, extending the approach to focus on the learning of the traditional STEM subjects, include the Arts, but also focusing on the use of drama to scaffold learners' understanding of STEM subjects (McGregor, 2017). This creative use of drama, often implemented in early years/primary education environments, provides opportunities for students to place themselves in scientific roles and encourages engagement in scientific activities (Kennedy & Tunnicliffe, 2022).

Many variations of the STEM acronym exist, such as STEMLE (Science, Technology, Engineering, Mathematics, Law and Economics), and STREM (Science, Technology, Robotics, Engineering and Multimedia), to name a few. STEM curriculum continues to evolve while educators around the world strive to provide application and problem-solving experiences to create more awareness of interdisciplinary opportunities for their students. While some educators advocate for the inclusion of the Arts and Humanities, others argue that STEM curriculum should include the history of science and highlight the contributions of women scientists as well as scientists of color to ensure gender equity and equality. The reality is that STEM education has a place in every aspect of schooling, as it incorporates engineering and technology concepts into the core subjects of math and science throughout the curriculum (see Chaps. 5 and 15 for more information about curriculum design in science and technology education as well as pedagogical content knowledge).

Although some scholars believe that adding an A, R or D is a dilution of STEM's focus and objectives, most agree that the iterations are symbolic reminders that STEM disciplines are enriched by other disciplines and facilitate the learning of all students. The recent COVID-19 pandemic revealed the potential critical thinking and problem-solving capabilities of STEM experts around the world. It also provided primary and secondary students with opportunities to develop and share their STEM solutions globally (Kennedy, 2021). STEM education has become more recognized as a societal solution, expanding across all content areas in a transdisciplinary fashion. Thus, this chapter will generally use the acronym STEM and recognize the contributions of all content areas subsumed within its applications.

Implementing STEM Education

As described earlier, defining STEM is the easy part. Implementing STEM education within the school setting is much more challenging. According to Bybee (2013), part of the problem is the general confusion about what STEM actually looks like in the classroom, since STEM education can take on various forms. It does not always incorporate all four STEM disciplines, and it is not always incorporated into project- or problem-based learning scenarios, also referred to as phenomenon-based learning in many countries (later described in Chaps. 18 and 20). However, nearly all STEM learning experiences have one thing in common—they provide students with opportunities to break down the artificial barriers between disciplines and enable students to better understand the connected nature of knowledge using critical skills, leading to success in the twenty-first century economy through applying the skills and knowledge that they have learned or are in the process of learning (Kennedy & Sundberg, 2020, p. 479). The basic tenet of STEM education aims to create critical thinkers, increase scientific literacy, and develop the next generation of innovators.

The 5E Model of Instruction (see Fig. 4.3) is recognized as one of the best processes by which educators can employ opportunities to personalize STEM learning for students of all ages. The five phases of the 5E Model, Engage, Explore, Explain, Elaborate and Evaluate, have "a 'common sense' value; it presents a natural process of learning" (Bybee, 2015, p. ix).

Based on the cognitive psychology, constructivist theory to learning, the five phases guide the learning process as students *engage* and focus on phenomena to make connections between past and present learning experiences; *explore* their environment using prior knowledge, generating new ideas through experimentation and trial and error to make sense of their surroundings; *explain* their observations

Fig. 4.3 Evidence-based practices: The 5E Model of Instruction. (Graphic used with permission from the San Diego County Office of Education, 2018, https://ngss.sdcoe.net/Evidence-Based-Practices/5E-Model-of-Instruction)

and understandings through their excitement and verbal explanations, and further construct deeper understandings as their peers and the adults around them provide additional information to help them make sense of their learning; and *elaborate* on their understanding through extended/enrichment activities. Students self-*evaluate* their learning and their teachers also evaluate their progress through informal formative assessments and formal summative assessments. See Chap. 14 for more information about educational psychology and its role in science education.

Conclusion

The debate of S.T.E.M versus STEM has generally subsided. Whichever form of STEM education we are referring to, whether it is STEAM, STREAM or others, STEM represents an educational philosophy centered on the integration of subjects (an integrated curriculum), and the ideal of building core competencies in twenty-first century skills (communication, collaboration, creativity, digital literacy, critical thinking and problem-solving) for every learner.

An examination of the research literature supports the interdisciplinary nature of STEM. That said, there is no clear configuration of interdisciplinary STEM derivatives or interdisciplinary variations. Regardless of the specific STEM strategy implemented, STEM education focuses on preparing all students to be problem-solvers and future leaders, workers and citizens who are flexible and can respond to new challenges locally and globally through innovation. STEM education increases student awareness of the technological world in which they live; how science, technology, engineering and mathematics support each other; how to creatively innovate and use new technologies as they become available; and how the technology decisions made directly impact their lives and the lives of others. "Twenty-first century students live in an interconnected, diverse, and rapidly changing world. Emerging economic, digital, cultural, demographic, and environmental forces are shaping young people's lives around the planet and increasing their intercultural encounters on a daily basis. This complex environment presents an opportunity and a challenge. Young people today must not only learn to participate in a more interconnected world but also appreciate and benefit from cultural differences. Developing a global and intercultural outlook is a process—a lifelong process—that education can shape" (OECD, 2018, p. 4).

Summary

This chapter explored STEM as a meta-discipline and discussed the origins and emergence of STEM education. The differences between S.T.E.M and STEM were defined, as were the many variations of the acronym. STEM education policy development continues to evolve in countries around the world, placing STEM literacy as a global priority since governments, business and industry, and educators place competence in STEM disciplines as essential to the global economy, competitiveness in the workforce, and in education. STEM education and research are necessary requirements to a nation's development, productivity, competitiveness, and societal wellbeing.

STEM is an interdisciplinary approach bringing the disciplines together to form a more applied science or meta-discipline, and plays a critical role in achieving the internationally agreed upon outcomes associated with the Sustainable Development Goals (SDGs). STEM education implementation models have evolved to encompass additional disciplines such as the Arts (STEAM), the Arts and drama (STEAMD), reading and writing (STREAM), design and design thinking (STEAMD), law and economics (STEMLE), robotics, engineering, and multimedia (STREM), and continue to evolve while educators around the world strive to provide application and problem-solving experiences to create more awareness of interdisciplinary STEM opportunities for their students.

The 5E Model of Instruction is recognized as one of the best processes by which educators can employ opportunities to personalize STEM learning for students of all ages. Through student-centered approaches such as Project-Based Learning, Problem-Based Learning, and Phenomenon-Based Learning, teachers can facilitate critical thinking through inquiry as they guide students to use their academic knowledge in real-world applications.

Recommended Resources

5 NSF-supported STEM education resources that are perfect for virtual learning https://beta.nsf.gov/science-matters/5-nsf-supported-stem-education-resources-are

21 Amazing STEM resources for teachers https://www.thetechedvocate.org/21-amazing-stem-resources-teachers/

10 great STEM sites for the classroom https://www.educationworld.com/a_lesson/great-stem-web-sites-students-classroom.shtml

References

Aina, J. K., & Akanbi, A. G. (2013). Perceived causes of students' low enrolment in science in secondary schools, Nigeria. *International Journal of Secondary Education, 1*(5), 18–22.

Al Murshidi, G. (2019). STEM education in the United Arab Emirates: Challenges and possibilities. *International Journal of Learning Teaching and Educational Research, 18*(12), 316–332.

Asia Society. (n.d.). *What is PISA and why does it matter? Global Cities Education Network.* https://asiasociety.org/global-cities-education-network/what-pisa-and-why-does-it-matter

Bybee, R. W. (2013). *The case for STEM education: Challenges and opportunities.* NSTA Press.

Bybee, R. W. (2015). *The BSCS 5E instructional model: Creating teachable moments.* NSTA Press. https://static.nsta.org/pdfs/samples/PB356Xweb.pdf

Freeman, B., Marginson, S., & Tytler, R. (2019, August). *An international view of STEM education.* Available at: https://www.researchgate.net/publication/335551705_An_international_view_of_STEM_education

Gonzalez, H. B., & Kuenzi, J. (2012). *Science, technology, engineering, and mathematics (STEM) education: A primer.* Congressional Research Service. http://www.stemedcoalition.org/wp-content/uploads/2010/05/STEM-Education-Primer.pdf

Henebery, B. (2020, December 11). *What makes an award-winning STEM program?* The Educator. https://www.theeducatoronline.com/k12/news/what-makes-an-awardwinning-stem-program/274741

Holbrook, J., Rannikmäe, M., & Soobard, R. (2020). STEAM education: A transdisciplinary teaching and learning approach. In B. Akpan & T. J. Kennedy (Eds.), *Science education in theory and practice: An introductory guide to learning theory* (pp. 465–477). Springer nature. https://doi.org/10.1007/978-3-030-43620-9

Kennedy, T. J. (2021, August 10). Student innovations related to COVID-19: The international engineering design challenge. *American Journal of Biomedical Science & Research. 13*(6), 592–595. https://biomedgrid.com/pdf/AJBSR.MS.ID.001921.pdf

Kennedy, T. J., & Odell, M. R. L. (2014). Engaging students in STEM education. *Science Education International, 25*(3), 246–258. https://eric.ed.gov/?id=EJ1044508

Kennedy, T. J., & Sundberg, C. W. (2020). 21st century skills. In B. Akpan & T. J. Kennedy (Eds.), chapter 32 *Science education in theory and practice: An introductory guide to learning theory* (pp. 479–496). Springer nature. https://doi.org/10.1007/978-3-030-43620-9

Kennedy, T. J., & Tunnicliffe, S. D. (2022) 'Introduction: The role of play and STEM in the early years'. In: *Play and STEM education in the early years: International policies and practices,* Tunnicliffe, S. D. & Kennedy, T. J. (Eds.), Chapter 1, pp. 3–37. Springer nature. https://link.springer.com/chapter/10.1007/978-3-030-99830-1_1

MacDonald, A., Hunter, J., Wise, K., & Fraser, S. (2019, July). STEM and STEAM and the spaces between: An overview of education agendas pertaining to "disciplinarity" across three Australian states. *Journal of Research in STEM Education, 5*(1), 75–92.

Marginson, S., Tytler, R., Freeman, B., & Roberts, K. (2013). *STEM: Country comparisons: International comparisons of science, technology, engineering and mathematics (STEM) education. Final report.* Australian Council of Learned Academies, Melbourne, Vic. https://www.researchgate.net/publication/305063165_STEM_country_comparisons_international_comparisons_of_science_technology_engineering_and_mathematics_STEM_education_Final

McGregor, D. (2017). *Using drama within a STEM context: Developing inquiry skills and appreciating what it is to be a scientist!* RADAR, Institutional repository of Oxford Brookes University. https://radar.brookes.ac.uk/radar/file/821f00c4-14bf-43f2-a4d8-b72f51323bae/1/mcgregor2017STEM.pdf

Milner, N., Ben-Zvi, R., & Hofstein, A. (1987). Variables that affect students' enrolment in science courses. *Research in science and technological education, 5*(2), 201.

NGSS Lead States. (2013). *'Appendices'. Next generation science standards: For States, by States, 2.* The National Academies Press.

OECD. (2018). *Preparing our youth for an inclusive and sustainable world.* Organisation for Economic Co-operation and Development. https://www.oecd.org/education/Global-competency-for-an-inclusive-world.pdf

OECD. (2019). *PISA 2018 results: Combined executive summaries.* Organisation for Economic Co-operation and Development. https://www.oecd.org/pisa/Combined_Executive_Summaries_PISA_2018.pdf

OECD. (n.d.). *What is PISA?* Organisation for Economic Co-operation and Development. https://www.oecd.org/pisa/

Sithole, A., Chiyaka, E., McCarthy, P., Mupinga, D., & Bucklein, B. (2017). Student attraction, persistence and retention in STEM programs: Successes and continuing challenges. *Higher Education Studies*, 7(1), 46–59. https://www.researchgate.net/publication/312474142_Student_Attraction_Persistence_and_Retention_in_STEM_Programs_Successes_and_Continuing_Challenges

UNDP. (2019). *Human development report 2019: Beyond income, beyond averages, beyond today: Inequalities in human development in the 21st century.* United Nations Development Programme. https://hdr.undp.org/system/files/documents//hdr2019pdf.pdf

UNESCO. (2017). *Education for sustainable development goals: Learning objectives.* https://unesdoc.unesco.org/ark:/48223/pf0000247444?utm_sq=gj34xbfn94

UNESCO International Bureau of Education. (2019, February). *Exploring STEM competencies for the 21st century, 30.* https://unesdoc.unesco.org/ark:/48223/pf0000368485

Zilberman, A., & Ice, L. (2021, January). *Employment projections for STEM occupations.* U.S. Bureau of Labor Statistics. https://www.bls.gov/opub/btn/volume-10/why-computer-occupations-are-behind-strong-stem-employment-growth.htm

Teresa J. Kennedy is a Professor of International STEM and Bilingual/ELL Education in the College of Education and Psychology at the University of Texas at Tyler. She holds appointments in the School of Education and the College of Engineering. She is a Past President of the International Council of Associations for Science Education, and currently serves on the ICASE Executive Committee, is a member of the UNESCO NGO Liaison Committee, and serves on the NSTA International Advisory Board. Her research interests include international comparative education, gender and equity issues in STEM, and content-based second language teaching and learning focused on STEM disciplines. Dr. Kennedy holds a PhD in Curriculum and Instruction from the University of Idaho.

Michael R. L. Odell is a Professor of STEM Education and holds the Sam and Celia Roosth Chair in the College of Education and Psychology. He holds appointments in the School of Education and the College of Engineering. He is the Co-Founder of the University Academy Laboratory Schools and serves on the School Board. He also provides oversight for the UA Curriculum. He is the Co-Director of the UTeach STEM Teacher Preparation program and the Co-Director of the EdD in School Improvement Program. His research interests are Education Policy, Sustainable Education, PBL, School Improvement, and STEM Education. Dr. Odell holds a PhD in Curriculum and Instruction from Indiana University.

Chapter 5
Curriculum Design in Science and Technology Education at International Level

Declan Kennedy

Abstract The signing of the Bologna Declaration by 48 countries has put the focus on learning outcomes as the foundation stone for curriculum design in science and technology education. However, there is still confusion about the use of terms such as *aims*, *objectives*, *learning intentions* and *competences* and how these terms are related to learning outcomes. This chapter discusses this relationship, as well as the implications for classroom teaching within a learning outcomes framework that stresses constructive alignment. It concludes by pointing out that, when implementing curricula in the classroom, only objectives and learning outcomes need to be used.

Keywords Aim · Objective · Learning outcome · Learning intention · Competence · Constructive alignment

Introduction

Since 1999, a quiet revolution has been taking place in curriculum design in education throughout the world. The seeds of this revolution were set in June 1999 when Ministers of Education of all European Union member states convened in Bologna, Italy, to formulate the Bologna Declaration. The overall aim of the Bologna Process was to improve the efficiency and effectiveness of education in Europe. One of the main features of this process was the need to improve the traditional ways of describing curricula and qualification structures. Prior to the Bologna Process, curricula were described in various ways in different countries, using terminology such as aims, objectives, goals and competences. However, these are vague terms and, as we shall discuss in this chapter, are open to interpretation. Hence, it was decided at

D. Kennedy (✉)
University College Cork, Cork, Ireland
e-mail: d.kennedy@ucc.ie

© The Author(s), under exclusive license to Springer Nature Switzerland AG 2023
B. Akpan et al. (eds.), *Contemporary Issues in Science and Technology Education*, Contemporary Trends and Issues in Science Education 56, https://doi.org/10.1007/978-3-031-24259-5_5

the meeting of Ministers of Education in Bologna that learning outcomes would be the common language for designing and developing curricula.

Learning Outcomes – The Common Language for Curricula

The signing of the Bologna Declaration in 1999 by 29 countries put the spotlight on the concept of using learning outcomes as the common language for teaching and learning in higher education. A total of 48 countries have now signed this declaration. In addition, the introduction of the European Qualifications Framework for Lifelong Learning (2008), based on learning outcomes, provided further momentum to teaching within a learning outcomes framework in primary, secondary and tertiary education. Thus, all programmes in primary, secondary and tertiary education in European Union countries, and many other jurisdictions, are now described in terms of learning outcomes. Similarly, all syllabi are now written in the form of learning outcomes in these countries. Hence, learning outcomes have become the common language to describe teaching, learning and assessment within 48 countries. Many other countries around the world have aligned the way in which they describe their national qualifications to the Bologna Declaration and the European Qualifications Framework in order to assist with international recognition of qualifications and student mobility.

One of the main reasons for embracing the concept of learning outcomes at international level is to bring clarity and coherence to the terminology used in education. Learning outcomes are clearly defined in the language of education and there is a common understanding of this term in the education literature. Some jurisdictions use terms such as 'learning intentions', 'success criteria' and 'competences' in curriculum design. The use of these terms in countries that have adopted a learning outcomes framework for teaching and learning has caused confusion among educators. It is hoped that this chapter will help to bring clarity to terms such as aim, objective, learning outcome, learning intention, success criteria and competence and assist educators to design curricula within a learning outcomes framework.

Aims and Objectives

The aim of a programme or scheme of work is a broad general statement of teaching intention, i.e., it indicates what the teacher intends to cover over an extended period of time. For example, one of the aims of the 3-year Junior Cycle science curriculum in Ireland is 'to develop students' evidence-based understanding of the natural world'. This is a very broad aim, which teachers hope to achieve over the 3 years of the programme.

On the other hand, teaching objectives (commonly referred to simply as 'objectives') tend to be more specific statements of teaching intention. For example, one of the objectives of a lesson could be to give students an appreciation of how they can contribute to sustainability through the recycling of materials.

When writing aims and objectives, we use terms such as:

- To give students an understanding of…
- To give students an appreciation of…
- To make students familiar with…
- To ensure that students know…
- To enable students to experience…
- To encourage students to…
- To provide students with the opportunity to…

What Is a Learning Outcome?

The concept of learning outcomes is very clearly defined in the literature (Morss & Murray, 2005; European Qualifications Framework, 2008; ECTS Users Guide, 2009; Kennedy et al., 2006).

Learning outcomes are statements of what a student should know, understand and be able to do after completion of a process of learning.

The 'process of learning' could be a lesson (or part of a lesson), or it could be a series of lessons, or an entire curriculum programme over several years.

In the classroom, learning outcomes can be used when describing individual lessons, topics within lessons or schemes of work over an extended period of time. In addition, learning outcomes can be used to design and develop entire curricula or programmes. These latter types of learning outcomes are often referred to as 'programme learning outcomes'. In third-level institutions, short courses are often referred to as modules, and the learning outcomes describing what students should know, understand and be able to do on completion of these short courses are commonly referred to as 'module learning outcomes'.

One of the main reasons that learning outcomes have become the international language of education is their flexibility across all types of courses at different levels of the educational system.

Benjamin Bloom (1913–1999) viewed learning as a process in which we build upon our former learning to develop more complex levels of understanding (Bloom et al., 1956). He carried out research into the development of classification of levels of thinking behaviours in the process of learning. He worked on drawing up levels of these thinking behaviours from the simple recall of facts at the lowest level up to evaluation at the highest level.

Knowledge is the foundation stone of Bloom's Taxonomy. Without knowledge, the other areas of Bloom's Taxonomy (understanding, application, analysis, synthesis and evaluation) cannot be achieved by our students. Bloom's Taxonomy in the cognitive ('knowing', 'thinking') domain is summarised in Fig. 5.1.

Fig. 5.1 Bloom proposed that our thinking can be divided into six increasingly complex levels, from the simple recall of facts at the lowest level to evaluation at the highest level. Knowledge is the foundation stone of Bloom's Taxonomy

Bloom devised the toolkit for writing learning outcomes. Since learning outcomes are statements that describe observable behaviour (what a student should be able to DO), the fundamental rule is that we must use action verbs when writing learning outcomes. An action verb describes the activity that the subject of a sentence is doing.

Examples of learning outcomes containing active verbs are:

- State the law of conservation of mass. *State* is the action verb. Explain why we need a blood system. *Explain* is the action verb. Evaluate the effects of climate change. *Evaluate* is the action verb.

A comprehensive list of action verbs for each level of Bloom's Taxonomy is given in Table 5.1.

The opposite type of verb to action verbs are stative verbs. Stative verbs describe a state. Some examples of stative verbs are *know, understand, appreciate, agree, imagine, wish* and *believe*.

Whilst action verbs are commonly used in writing learning outcomes, stative verbs should never be used in writing learning outcomes. Common mistakes used in writing learning outcomes are:

- Using the term 'understand'. Instead of this term, ask the students to show their understanding by using a learning outcome that contains action verbs such as *explain, discuss, illustrate* or *solve.*
- Using the term 'appreciate' in the cognitive domain. Instead of this term, ask the students to show their appreciation of a specific concept by asking them to *evaluate, discuss, outline* or *summarise.*

Since aims and objectives are written by the teacher; they are often associated with the 'teacher-centred' approach to teaching and learning. However, since learning outcomes focus on what the student can DO, they are often associated with the 'student-centred' approach to teaching and learning (see Fig. 5.2).

Table 5.1 Examples of action verbs that may be used to write learning outcomes

Knowledge	Comprehension	Application	Analysis	Synthesis	Evaluation
Arrange, collect, define, describe, duplicate, enumerate, examine, find, identify, label, list, locate, memorise,name, order, outline, present, quote, recall, recognise, recollect, record, recount, relate, repeat, reproduce, show, state, tabulate, tell.	Associate, change, clarify, classify, construct, contrast, convert, decode, defend, describe, differentiate, discriminate, discuss, distinguish, estimate, explain, express, extend, generalise, identify, illustrate, indicate, infer, interpret, locate, predict, recognise, report, restate, review, select, solve, translate.	Apply, assess, calculate, change, choose, complete, compute, construct, demonstrate, develop, design, discover, dramatise, employ, examine, experiment, find, illustrate, interpret, manipulate, modify, operate, organise, practise, predict, prepare, produce, relate, schedule, select, show, sketch, solve, transfer, use.	Analyse, appraise, arrange, break down, calculate, categorise, classify, compare, connect, contrast, criticise, debate, deduce, determine, differentiate, discriminate, distinguish, divide, examine, experiment, identify, illustrate, infer, inspect, investigate, order, outline, point out, question, recognise, relate, separate, solve, sub-divide, test.	Argue, arrange, assemble, categorise, collect, combine, compile, compose, construct, create, develop, design, devise, establish, explain, formulate, generate, generalise, infer, integrate, invent, make, manage, modify, organise, originate, plan, prepare, propose, rearrange, reconstruct, relate, reorganise, revise, rewrite, set up, summarise.	Appraise, argue, ascertain, assess, attach, choose, compare, conclude, contrast, convince, criticise, decide, defend, discriminate, explain, evaluate, interpret, judge, justify, measure, predict, rate, recommend, relate, resolve, revise, score, summarise, support, validate, value.

In the past, all curricula were described only in terms of aims and objectives. Since words such as *know* and *understand* are rather vague, very often it was not clear what exactly was expected of students. The great advantage of having to write learning outcomes is that, when we engage in the process of writing learning outcomes, this process forces us to think about our teaching, learning and assessment. It is rather like looking through the eyepiece of a microscope and observing a fuzzy image. However, focusing the microscope, which can be compared to implementing a learning outcomes framework in our teaching, displays a far clearer image.

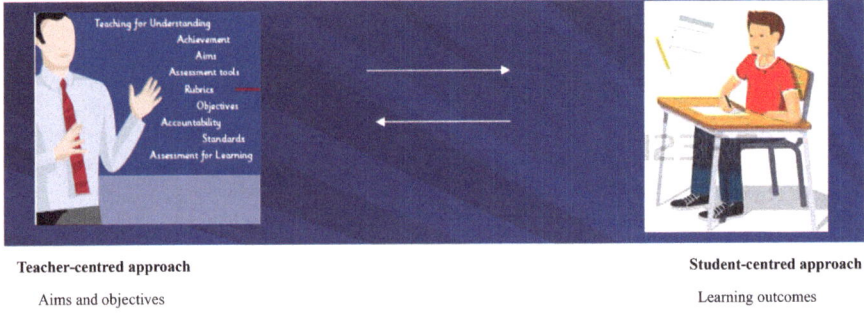

Teacher-centred approach	**Student-centred approach**
Aims and objectives	Learning outcomes

Fig. 5.2 The teacher-centred and student-centred approaches are inextricably linked in a dynamic equilibrium in the classroom as shown by the arrows pointing in both directions

International Best Practice in Curriculum Design

One does not have to look far to find exemplars of good practice in curriculum design using a learning outcomes framework. One of the most prestigious awarding bodies at international level is the OCR (Oxford, Cambridge and Royal Society of Arts) examination board in England. This awarding body develops curricula in over 40 subjects and offers over 450 vocational qualifications at national and international level. Copies of the GCSE science curricula taken at age 16 by students in England and Wales may be found online (OCR, 2022a, b, c). Even a cursory glance at some pages from these curricula (see Fig. 5.3) shows that learning outcomes are the starting points for each topic on the curriculum.

It is clear from the above examples that learning outcomes are the foundation stone for curriculum design. Note also that, where necessary, in the second column clarification is given on the depth of treatment that should be included in one's teaching to help students achieve the learning outcome. The remaining columns are used to provide guidance to the teacher on the mathematical and scientific skills that should be developed, as well as suggestions for practical work to enable the learning outcome to be achieved by students.

Constructive Alignment

In designing and developing any curriculum, there must be a clear linking of learning outcomes to teaching and learning activities and also to assessment. Biggs (2003) coined the phrase 'constructive alignment' to explain this concept. Alignment refers to what the teacher does in helping to support the learning activities to achieve the learning outcomes. The teaching methods and the assessment must be aligned to the learning activities designed to achieve the learning outcomes. Aligning the assessment with the learning outcomes means that students know how their achievements will be measured.

Topic content		Opportunities to cover:		Practical suggestions
Learning outcomes	To include	Maths	Working scientifically	
B2.1a explain how substances are transported into and out of cells through diffusion, osmosis and active transport	examples of substances moved, direction of movement, concentration gradients and use of the term water potential (no mathematical use of water potential required)	M1c, M1d	WS2a, WS2b, WS2c, WS2d	Observation of osmosis in plant cells using a light microscope. Investigation of 'creaming yeast' to show osmosis. (PAG B6, PAG B8) Investigation into changes in mass of vegetable chips when placed in sucrose/salt concentrations of varying concentrations. (PAG B6, PAG B8)
B2.1b describe the process of mitosis in growth, including the cell cycle	the stages of the cell cycle as cell growth, DNA replication, more cell growth, movement of chromosomes		WS2a, WS2b, WS2c, WS2d	Modelling of mitosis using everyday objects e.g. shoes, socks etc. Observation of mitosis in stained root tip cells. (PAG B1, PAG B6, PAG B7)
B2.1c explain the importance of cell differentiation	the production of specialised cells allowing organisms to become more efficient and examples of specialised cells		WS2a, WS2b, WS2c, WS2d	Examination of a range of specialised cells using a light microscope. (PAG B1)

Topic content		Opportunities to cover:		Practical suggestions
Learning outcomes	To include	Maths	Working scientifically	
C4.1a recall the simple properties of Groups 1, 7 and 0	physical and chemical properties		WS1.2a, WS1.4a, WS1.4c	Displacement reactions of halogens with halides. (PAG C1)
C4.1b explain how observed simple properties of Groups 1, 7 and 0 depend on the outer shell of electrons of the atoms and predict properties from given trends down the groups	ease of electron gain or loss; physical and chemical properties			
C4.1c ☑ recall the general properties of transition metals and their compounds and exemplify these by reference to a small number of transition metals	melting point, density, reactivity, formation of coloured ions with different charges and uses as catalysts		WS1.4a	Investigation of transition metals. (PAG C1, PAG C5, PAG C8)

Topic content		Opportunities to cover:		Practical suggestions
Learning outcomes	To include	Maths	Working scientifically	
P3.1a describe that charge is a property of all matter and that there are positive and negative charges	the understanding that in most bodies there are an equal number of positive and negative charges resulting in the body having zero net charge		WS1.1b, WS1.1e, WS1.2a, WS1.3e, WS2a	Use of charged rods to repel or attract one another. Use of a charged rod to deflect water or pick up paper. Discussion of why charged balloons are attracted to walls.
P3.1b describe the production of static electricity, and sparking, by rubbing surfaces, and evidence that charged objects exert forces of attraction or repulsion on one another when not in contact	the understanding that static charge only builds up on insulators		WS1.1b, WS1.1e, WS1.2a, WS1.3e	Use of a Van de Graaff generator.
P3.1c explain how transfer of electrons between objects can explain the phenomena of static electricity			WS1.1b, WS1.3e, WS1.3f, WS2a	Use of the gold leaf electroscope and a charged rod to observe and discuss behaviour.
P3.1d ☑ explain the concept of an electric field and how it helps to explain the phenomena of static electricity	how electric fields relate to the forces of attraction and repulsion	M5b	WS1.3e	Demonstration of semolina on castor oil to show electric fields.

Fig. 5.3 Some screenshots from the OCR biology, chemistry and physics GCSE syllabi developed using a learning outcomes framework

Schuell emphasised the importance of linking learning outcomes to teaching and learning activities:

If students are to learn desired outcomes in a reasonably effective manner, then the teacher's fundamental task is to get students to engage in learning activities that are likely to

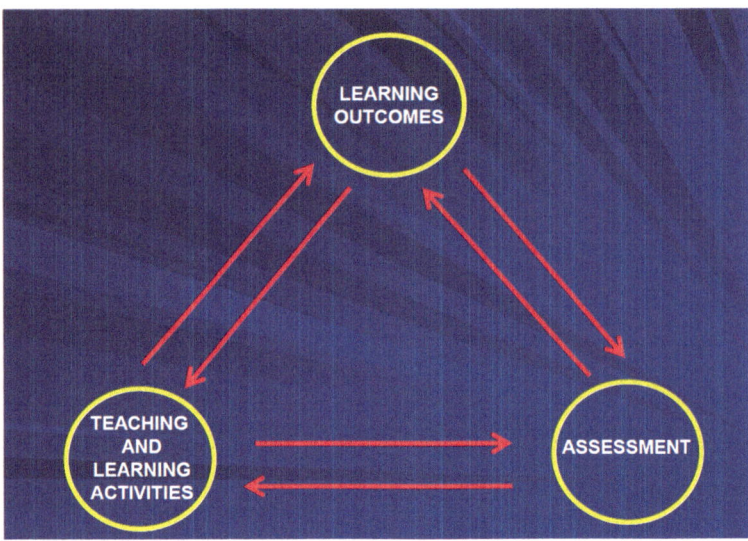

Fig. 5.4 Constructive alignment involves co-ordination between the learning outcomes, teaching and learning activities and assessment

> *result in their achieving those outcomes…It is helpful to remember that what the student does is actually more important in determining what is learned than what the teacher does* (Shuell, 1986).

Constructive alignment emphasises the importance of what the teacher does in helping to support the learning activities to achieve the learning outcomes. The teaching methods and the assessment are aligned to the learning activities designed to achieve the learning outcomes. Aligning the assessment with the learning outcomes means that students have a clear vision of how their achievements will be measured.

Thus, constructive alignment is the deliberate linking within curricula of learning outcomes, teaching and learning activities and assessment. Assessment must be designed such that students are able to demonstrate that they have met the learning outcomes. Morss and Murray (2005) described constructive alignment as simply a fancy name for 'joining up the dots'. The author finds Fig. 5.4 useful when explaining constructive alignment to his students.

In short, constructive alignment is a key ingredient in the design of any curriculum within a learning outcomes framework.

Teaching and Learning Within a Learning Outcomes Framework

In order to illustrate the key role of learning outcomes in modern curriculum design, let us consider a case study. In the BSc Science Education degree programme of University College Cork, Ireland, student teachers are trained how to teach within a learning outcomes framework. At the beginning of each lesson taught to their pupils, the trainee teacher shares and discusses the teaching objectives with the pupils. Then as the lesson proceeds through the introductory and development phases, the learning outcomes are shared with the pupils and strategies are included in the lesson plan (questioning, worksheets, group work, etc.) to help teacher and pupils assess if the learning outcomes are being achieved. This is an example of formative assessment in action. In the summary and recapitulation phase of the lesson plan, the full list of learning outcomes achieved by the pupils in the lesson is discussed with the class. Finally, homework is allocated to test the extent to which each pupil has achieved the learning outcomes (summative assessment).

The key to teaching within a learning outcomes framework involves linking each learning outcome with an appropriate form of assessment. The active verb in the learning outcome is always a great clue to devising the appropriate form of assessment. If the assessment is properly chosen, it will provide evidence that the students have achieved the learning outcomes during the lesson.

Each student teacher on the BSc Science Education degree is visited a number of times during the year by a teaching placement tutor who supervises and supports the student teacher. All of these teaching placement tutors are highly experienced science teachers who have spent many years teaching science at secondary school level. Common types of question that are often asked after an observed lesson include:

1. Do you think that the pupils achieved the learning outcomes of the lesson?
2. What evidence do you have that pupils achieved these learning outcomes?

An example of the type of evidence that is looked for is summarised in Table 5.2.

Every time the teacher writes a learning outcome, he or she should always have one question at the back of their minds. This question is: 'How can I assess this learning outcome?', i.e., how do I know if my students have achieved the learning outcome and how can I measure the extent to which each student has achieved the learning outcome? Since learning can often be invisible, we must make it visible by asking the students to DO something to make the learning visible.

Table 5.2 Examples of evidence gathered during a lesson (formative assessment) to check if pupils have achieved the learning outcomes

Learning Outcome	Assessment (evidence of level of achievement of the learning outcomes)
List examples of everyday acids and bases.	Students <u>completed</u> the worksheet about acids and bases found in the home
Classify substances as acidic, basic or neutral.	In the group work, students were able to <u>separate</u> the acids, bases and neutral substances into different categories.
Test a variety of solutions with litmus and classify these as acidic, basic or neutral.	In the practical work activity, students were able to <u>interpret</u> the colour changes of the litmus paper and hence were able to <u>divide</u> the given substances into three different groups.
Investigate the pH of a variety of materials.	Students were able to <u>use</u> the pH paper and colour chart to conclude if the solutions were highly acidic, mildly acidic, neutral, mildly basic or highly basic.

In short, teaching within a learning outcomes framework involves making the learning outcomes the driving force for teaching, learning and assessment strategies.

Success Criteria – What Does This Term Mean?

The Australian Institute for Teaching and School Leadership (AITSL) defines success criteria as *'the measures used to determine whether, and how well, learners have met the learning intentions'* (AITSL, n.d.). The term 'success criteria' is simply another term for the word 'assessment'. There is no need to use the term 'success criteria' when teaching within a learning outcomes framework. When teaching within this framework, we simply assess if the learning outcomes have been achieved by each student and also assess the extent to which each learning outcome has been achieved.

When training science teachers on the BSc. Ed programme, we do not used the term 'success criteria'. We simply ask the student teacher to provide the evidence that the learning outcomes have been achieved by students.

What Are Learning Intentions?

The Australian Institute for Teaching and School Leadership (AITSL) defines learning intentions as follows: *'Learning intentions are descriptions of what learners should know, understand and be able to do by the end of a learning period or unit'* (AISTL, n.d.).

This is the same definition as the definition of learning outcomes. The Ministry of Education in New Zealand clearly states in published documentation that learning intentions and learning outcomes mean the same thing (Ministry of Education New Zealand, n.d.). The interpretation of learning intentions being identical to learning outcomes is also made in many other publications (Wu & Goff, 2021; SCPI, 2007).

The term *learning intention* is commonly used in publications on formative assessment (Black & Wiliam, 2009; Bennett, 2011; Andersson & Palm, 2017). In general, it is clear from these research publications on formative assessment that the term *learning intention* may be used interchangeably with the term *learning outcome*. For example, the learning intention *'Draw the skeletal system, label the parts and identify their functions'* in the publication on formative assessment by Wylie and Lyon (2015) is a good example of a learning outcome.

As discussed earlier, the rules for writing learning outcomes using Bloom's Taxonomy are clearly outlined in the literature. However, there is a lack of consistency in the literature when writing learning intentions. For example, Robertson (2020) gives the statement 'Understand what deforestation is' as an example of a learning intention. This author appears to be equating learning intentions with the concept of an objective rather than a learning outcome. Wu and Goff (2021) also adopt a loose interpretation of the term 'learning intentions' and give the statement 'We're going to carry on talking about shapes' as an example of a learning intention.

This lack of consistency in the writing of learning intentions may be one of the contributing factors causing difficulty for teachers in writing learning intentions, as reported by Webb (2010).

Thus, it is not surprising that the lack of a coherent approach to writing learning intentions is causing confusion among teachers. In short, whilst the rules for writing learning outcomes are very clear in the literature, there is a lack of consistency in the interpretation of the term 'learning intentions' and also in the writing of learning intentions.

The Use of Competence and Competency in Curriculum Design

In some countries, the term competence is used in curriculum design. There is considerable confusion in the literature with regard to the meaning of the term *competence* and the relationship between competences (also written as competencies) and learning outcomes. Some authors interpret the terms 'competence' and 'competency' to have identical meaning and others define the terms differently. The situation is neatly summarised by van der Klink and Boon (2003), who attempt to trace the different interpretations of the concept of competence within the educational systems of various countries:

> There is considerable confusion about what competency actually means…First, differences can be observed between nations along the lines of different national educational policies and different types of relations between education and the labour market, many of which have an historic origin. In the British approach it refers to the ability to meet the performance standards for functions and professions such as those developed for National Vocational Qualifications (NVQs) in the UK. In the USA, competencies refer to the skills, knowledge and characteristics of persons, that is traits, motives and self-concept, which contribute to performance excellence. These differences are evident even in the words themselves: competences (UK) and competencies (USA). To put it simply: competences refer to work and its achievement; competencies concern the people who do the work… More than in the UK or the USA, the German perspective stresses a holistic view of competency. It is not just a random collection of skills and knowledge. Competencies are defined as integrated action programmes that enable individuals to perform adequately in various job contexts within a specific profession (Van der Klink & Boon, 2003).

A review of the literature in the area of competences shows that there is no single definition of the term 'competence'. Descriptions of the term range from that of a broad overarching attribute to that of a very specific task. The situation is nicely summarised by Brown (1994) when discussing the use of the term in the context of managerial competence:

One of the reasons for the debate about the usefulness of managerial competence may be the soft focus and blurred edges of the term "competence". Social science has the habit of taking a word from our common vocabulary and altering the meaning by its adoption as a technical or academic term. This process is still happening to "competence" and a common consensus has yet to be established as to what the word should mean when used in management applications (Brown, 1994).

The above conclusion is echoed by van der Klink and Boon (2003) when discussing the fuzzy concept of competences:

The fact that the concept of competencies serves as a remedy for solving rather different problems probably has to do with its diffuse nature. It is actually an ill-defined concept with no clear content, thus allowing ample interpretations. This major vagueness is partly caused by the application of the concept in various countries, different settings and for different purposes. Its vagueness is probably at the same time the explanation for its prominent status today but it makes it difficult to use the concept as a sound cornerstone for designing HRD [Human Resource Development] and educational practices (Van der Klink & Boon, 2003).

The confusion surrounding the use of the term 'competence' is in contrast with the clear definition of the concept of a learning outcome found in the literature (Morss & Murray, 2005; European Qualifications Framework, 2008; ECTS Users Guide, 2009; Kennedy, 2007).

Given the considerable confusion in the literature, if the term 'competence' is used in curriculum design, then its meaning needs to be clearly defined for the context in which it is being used. It is obvious from the literature that, within certain professions, the term 'competence' has a shared meaning, e.g., in dentistry, competence refers to psychomotor skill performance and understanding of what is being done and is supported by professional values (Chambers, 1994). Hence, there is no problem with using the concept of competence, since there is a common understanding of its meaning among the members of that profession. The problem arises when the term 'competence' is used in a general context without defining what it means.

Since there is not a common understanding of the term 'competence', learning outcomes have become more commonly used than competences when describing what students are expected to know, understand and/or be able to demonstrate at the end of a module or programme. The 'fuzziness' of competences disappears in the clarity of learning outcomes!

Conclusions and Recommendations

The Bologna Process has placed the focus on the role of learning outcomes in curriculum design and development. This focus was led by the 48 countries that signed up to the Bologna Declaration, but has now spread throughout the world as countries realise the importance of international recognition and student mobility. Learning outcomes serve as the foundation stone for developing and designing curricula at primary, secondary and tertiary level.

In designing and developing any curriculum, there must be a clear linking of learning outcomes to teaching and learning activities and also to assessment. In other words, constructive alignment must always exist within the curriculum.

The use of competences in curriculum design is confusing due to the 'fuzzy' nature of the concept. To ensure clarity of meaning, it is recommended to write competences using the vocabulary of learning outcomes, i.e., express the required competence in terms of the students achieving specific programme learning outcomes, or module learning outcomes.

The definition of learning outcomes in the literature at international level and the definition of learning intentions by government organisations in countries such as Australia and New Zealand are identical. Hence, learning outcomes and learning intentions may be used interchangeably. Whilst the rules for writing learning outcomes are clearly defined, there is lack of consistency of practice in the literature when writing learning intentions. Whilst most authors write learning intentions using the language of learning outcomes, some authors write learning intentions using the language of objectives.

When teaching in the classroom within a learning outcomes framework, the use of two simple terms is recommended in lesson planning: objectives and learning outcomes. Everything that is needed in a lesson can be covered by sharing the objectives of the lesson and the learning outcomes with our students.

For countries that are aligned to the Bologna Declaration and the European Qualifications Framework, it is recommended that learning outcomes alone be used in the classroom for lesson planning. Using **both** the terms 'learning outcomes' and 'learning intentions' in lesson preparation is unnecessary and confusing.

Summary

In this chapter, I have discussed the concepts of aims, objectives and learning outcomes. I have given an overview of international best practice in curriculum design and how constructive alignment is embedded within curriculum design. I have also discussed key areas of teaching and learning within a learning outcomes framework. Finally, we have clarified the role of competence and competency in curriculum design.

Recommended Resources

Resources for curriculum development http://essentialschools.org/horace-issues/resources-for-curriculum-development/

Resources for developing curriculum and teaching materials https://www.sil.org/literacy-education/resources-developing-curriculum-and-teaching-materials

References

AITSL (Australian Institute for Teaching and School Leadership). (n.d.). *Learning Intentions and Success Criteria*. Available at: https://www.aitsl.edu.au/docs/default-source/feedback/aitsl-learning-intentions-and-success-criteria-strategy.pdf?sfvrsn=382dec3c_2

Andersson, C., & Palm, T. (2017). Characteristics of improved formative assessment practice. *Education Inquiry, 8*(2), 104–122.

Bennett, R. E. (2011). Formative assessment: A critical review. *Assessment in Education: Principles, Policy & Practice, 18*(1), 5–25.

Biggs, J. (2003, April 13–17). Aligning teaching and assessing to course objectives. In: *Teaching and learning in higher education: New trends and innovations*. University of Aveiro.

Black, P., & Wiliam, D. (2009). Developing the theory of formative assessment. *Educational Assessment Evaluation and Accountability, 21*, 5–31. https://www.researchgate.net/publication/225590759_Developing_the_theory_of_formative_assessment

Bloom, B. S., Engelhart, M. D., Furst, E. J., Hill, W., & Krathwohl, D. (1956). *Taxonomy of educational objectives. Volume I: The cognitive domain*. McKay.

Brown, R. B. (1994). Reframing the competency debate: Management knowledge and meta-competence in graduate education. *Management Learning, 25*(2), 289–299.

Chambers, D. W. (1994). Competencies: A new view of becoming a dentist. *Journal of Dental Education, 58*, 342–345.

ECTS Users Guide. (2009). Available at: http://www.ehea.info/media.ehea.info/file/ECTS_Guide/77/4/ects-guide_en_595774.pdf

European Qualifications Framework. (2008). Available at: https://europa.eu/europass/en/european-qualifications-framework-eqf

Kennedy, D. (2007). *Writing and using learning outcomes – A practical guide. Quality promotion unit*. University College Cork. Available from: cora.ucc.ie/handle/10468/1613

Kennedy, D., Hyland, A., & Ryan, N. (2006). Writing and using learning outcomes. *Bologna Handbook, Implementing Bologna in your Institution, C3*(4–1), 1–30.

Ministry of Education New Zealand. (n.d.). *Getting to grips with learning intentions and success criteria*. An online workshop. Available at: https://slideplayer.com/slide/5982278/

Morss, K., & Murray, R. (2005). *Teaching at university*. Sage Publications. ISBN 1412902975.

OCR Gateway Biology. (2022a). https://www.ocr.org.uk/qualifications/gcse/gateway-science-suite-biology-a-j247-from-2016/

OCR Gateway Chemistry. (2022b). https://www.ocr.org.uk/qualifications/gcse/gateway-science-suite-chemistry-a-j248-from-2016/

OCR Gateway Physics. (2022c). https://www.ocr.org.uk/qualifications/gcse/gateway-science-suite-physics-a-j249-from-2016/

Robertson, B. (2020). *The teaching delusion*. Available at: https://theteachingdelusion.com/2019/11/27/a-five-minute-guide-to-learning-intention-success-criteria/

SDPI. (2007). *School development planning initiative. Assessment for learning – Improving teaching and learning in our school*. Available at https://slidetodoc.com/school-development-planning-initiative-an-initiative-for-schools-5/

Shuell, T. J. (1986). Cognitive conceptions of learning. *Review of Educational Research, 56*, 411–436.

Van der Klink, M., & Boon, J. (2003). Competencies: The triumph of a fuzzy concept. *International Journal of Human Resources Development and Management*, Inderscience Enterprises Ltd, *3*(2), 125–137.

Webb, M. (2010). Beginning teacher education and collaborative formative e-assessment. *Assessment and Evaluation in Higher Education, 35*(5), 597–618.

Wu, B., & Goff, W. (2021). Learning intentions: A missing link to intentional teaching? Towards an integrated pedagogical framework. *Early Years Research Journal*, 1–15. https://doi.org/10.1080/09575146.2021.1965099

Wylie, A., & Lyon, C. (2015). The fidelity of formative assessment implementation: Issues of breadth and quality. *Assessment in Education: Principles, Policy and Practice, 22*(1), 140–160.

Declan Kennedy graduated from University College Cork Ireland with a BSc in Chemistry (major) and Mathematical Physics (minor), a Postgraduate Diploma in Education and an MSc in X-ray crystallography. He subsequently studied at the University of York, England and graduated with a Master's Degree in Education and a PhD in Education. He spent over 20 years teaching science at secondary school level, was appointed lecturer in science education at University College Cork and was subsequently promoted to senior lecturer. He regularly speaks at conferences in Ireland and abroad and organises many workshops in the area of education, curriculum planning and assessment. To date, he has been invited to give lectures and workshops in 34 countries and has been closely involved in helping many countries to fulfil the requirements of the Bologna Process in terms of curriculum frameworks, teaching and learning activities, learning outcomes and assessment.

Chapter 6
Assessment and Evaluation in Science and Technology Education

Bulent Çavaş, Pınar Çavaş, and Şengül Anagün

Abstract This chapter is focused upon the assessment and evaluation dimension in science and technology education. In this chapter, (i) the definition of assessment and evaluation is offered and the explanations of assessment and evaluation in science and technology education within (ii) the scope of twenty-first century skills are given. The chapter provides detailed information on (iii) commonly-used assessment and evaluation tools in science and technology education, and how these tools can be used in learning and teaching environments. The chapter also includes (iv) Web 2.0 technologies that can be used for assessment and evaluation purposes, the introduction of commonly-used Web 2.0 tools in this field, and their capacities and limitations. The last part of the chapter includes (v) discussions about assessment and evaluation in science and technology education and also offers recommendations. The final section also presents readers with examples of books and journal articles that provide more detailed information for assessment and evaluation in science and technology education.

Keywords Assessment · Evaluation · Web 2.0 tools · Science and technology education

B. Çavaş (✉)
Faculty of Education, Dokuz Eylül University, Buca, Izmir, Türkiye
e-mail: bulent.cavas@deu.edu.tr

P. Çavaş
Faculty of Education, Ege University, Bornova, Izmir, Türkiye
e-mail: pinar.cavas@ege.edu.tr

Ş. Anagün
Eskişehir Osmangazi University, Faculty of Education, Eskişehir, Türkiye
e-mail: ssanagun@ogu.edu.tr

© The Author(s), under exclusive license to Springer Nature Switzerland AG 2023
B. Akpan et al. (eds.), *Contemporary Issues in Science and Technology Education*, Contemporary Trends and Issues in Science Education 56, https://doi.org/10.1007/978-3-031-24259-5_6

Introduction

Global changes in science and technology in the twenty-first century have changed the meaning, purpose and methods of education. The main purpose of education is to raise individuals who can adapt to today's developing world and have the skills to play a dynamic role in society's social and economic situations (MoNE, 2017). It is clear that science and technology education plays an important role in the twenty-first century, where scientific knowledge is increasing exponentially, technological innovations are advancing at a rapid pace, and their impact on societies is increasing (for more information about science, technology and society, please see Chaps. 10, 11, 12, and 13). This situation forces all societies to continuously strive to improve the quality of science education. The quality of science education is controlled by assessment and evaluation studies.

The main purpose of this chapter is to explain assessment and evaluation processes in science and technology education. First, the concept of assessment and evaluation will be explained, then the twenty-first century's understanding of assessment and evaluation will be explored. The chapter will end with examples of alternative assessment and evaluation tools used in science and technology education and the use of Web 2.0 tools in assessment.

What Is Assessment and Evaluation?

It is quite clear that science and science education is at the center of the education systems of all countries. Today, scientific knowledge has never been more critical to make sense of what is happening around us. Understanding scientific studies is extremely important in order to grasp what is happening in current life, to use technology effectively, or to make informed decisions about one's life (NRC, 2013). This situation makes the assessment and evaluation of the outputs achieved in science and technology education more important. In this context, it will be useful to explain assessment and evaluation as a concept before moving on to the characteristics of measurement and evaluation in science education.

'Assessment' and 'evaluation' are often used synonymously in daily life. However, these two concepts are different from each other. 'Assessment' is used to judge any learning or performance, while 'evaluation' means measuring academic effort in all its aspects (Martin & Collins, 2011). Evaluation is carried out with the aim of determining the value of any learning based on certain criteria (Boonchutima & Pinyopornpanich, 2013). Assessment can be expressed as continuous and systematic measurements taken to review the learner's strengths and weaknesses, their developments based on data and evidence, and to provide the academic support that they need (Yambi, 2018). On the other hand, evaluation confirms and judges the performance or learning outcome. Therefore, the main difference concerning the two concepts is that assessment is focused towards progress and evaluation is directed to consequence. In other words, evaluation is the final step in assessing the quality of a completed process (Yambi, 2018, cited in Mubayrik, 2020).

Yambi (2018) explained the main differences between assessment and evaluation as follows:

1. Assessment can be considered as a data collection and review process for achievement. Evaluation refers to the process of judging grades or scores based on standard criteria.
2. Assessment is diagnostic as it identifies weaknesses that need improvement. Evaluation is judgmental as it gives the student a total score as a result of all that they have done.
3. Assessment works towards an answer about learning to improve the performance. Evaluation decides if the standards are met or not.
4. The goal of assessment is formative, to improve the performance during the process. On the other hand, evaluation is summative, since it is performed after the process has been finalized to judge the quality of learning.
5. Assessment aims at the process, while evaluation tends towards the outcome.
6. Assessment responses rely on considerations of strong and weak points. However, in evaluation, it depends on the level of outcome compared to pre-set criteria.

Twenty-First Century Skills and the Meaning of Assessment

The twenty-first century is witnessing important developments in the field of education as well as in every other field (for more information about 21st skills, please see chapters at section III). In order to contribute to society in the twenty-first century, individuals should not only learn the content of a field of knowledge, but also have the innovation, technology and career skills required for business life. Science and technology teaching is one of the subjects most affected by these changes and circumstances make it necessary to enhance new perspectives. Therefore, science and technology education will be concentrated in competencies desirable in the twenty-first century, such as critical thinking, creativity, communication and collaboration, which are known as the Four C competencies. These four competencies are, together, generating the need for new forms of learning and contributing the guiding principles required to support twenty-first century learning practices. Becoming competent in any field means acquiring knowledge and skills about the subject area. Competency also requires being able to apply knowledge to the solution of different kinds of problems. Teaching about subject matter, scientific process skills and reasoning skills is vital for science instruction. There is a close association between teaching and assessment. All types of learning need to be evaluated. Therefore, the 'competencies' developed in the science and technology teaching-learning process should be evaluated in an appropriate way.

One new way of assessment for twenty-first century outcomes is authentic assessment. Authentic assessment is a concept used to express different assessment tools used in addition to standard paper-pencil tests. These types of assessments are more sensitive and proceed in several forms. Instead of focusing on the right answers to the questions, they underline and attempt to describe how individuals process information, build new information and resolve problems. An alternative assessment approach makes available the vehicle through which various types of assessment tools can be unified and used to define an individual's development, and it is to this that we now turn.

Alternative Assessment Tools in Science and Technology Education

Alternative assessment is an umbrella term that refers to assessment methods for providing alternatives to the traditional paper-and-pencil assessments. In the literature, 'authentic assessment', 'portfolio assessment' and 'performance assessment' are terms sometimes used interchangeably with 'alternative assessment' (Chittenden, 1991; Shanklin & Conrad, 1991; Gipps & Stobart, 2003). Unlike traditional assessment, alternative assessment actively requires students to participate in the process of *'what is taught, how it is taught, and how it is evaluated'* (Kreisman, Knoll & Melchior, 1995, p.114). According to Gipps and Stobart (2003), this assessment is not purely the usage of alternative forms of assessment, but is also an alternative use of assessment as part of the learning process. Students should actively set goals, perform tasks, create products and use their metacognitive skills. While alternative-based approaches are based on cognitive (or constructivist) theory, multiple-choice standardized tests were readily adopted by behaviorist views of 'knowledge in pieces'.

Alternative assessment, which can be thought of as a complementary component for students with different learning styles, offers students a way to prepare their answers in a way that traditional assessment does not (Stiggins, 1994). The alternative assessment approach focuses more on processes than results and uses multiple different techniques. When treated as a versatile assessment, the points to be considered can be listed as:

- Assessment should be long-term.
- Assessment should include many skills and types of intelligence.
- Individual and group assessments should be made.
- Assessment should focus on both the product and the process.
- Multiple data collection techniques should be used in the evaluation (Cepni & Ayvaci, 2016).

In the literature, it is seen that there are different alternative assessment methods, techniques and tools used in science education. Some of them are explained below.

Performance Assessment

In this assessment approach, students are asked to demonstrate what they have learned from the course and to what extent, and to show what they know through the open-ended tasks given by their teachers. McTighe and Ferrara (1998) mentioned that performance-based assessments should include students' products (essay, research paper, portfolio, science project, model, etc.), students' performance (oral presentation, science lab demonstration, debate and teach-a-lesson) and process-focused assessment (oral questioning, observation, interview, conference, learning log, etc.). Performance-based assessments are about showing how students use their knowledge and skills in different situations and events, rather than revealing the knowledge that they have acquired through memorization. In this type of assessment, there is no single correct answer for students to give. Alternative solution proposals put forward by the students regarding the performance status that they are asked to show are more valuable. Such evaluations are determined by the judgements to be made in line with the determined criteria.

Rubrics

A rubric is known as a popular scoring tool around the world. The term 'rubric' has its origins in the Latin word *rubrica,* meaning 'red earth used to mark something of significance'. Today, educators use rubric to communicate the important qualities in a product or performance (McTighe & Ferrara, 1998, p.21). For the development of rubrics, there should be a fixed measurement scale and a list of criteria that can work in accordance with these measurements. For each criterion, a score must be determined. For example, rubrics can be used effectively in scoring the activities developed by the groups in the whole class for the science fair. Rubrics can be used as assessment tools in two different ways: holistic and analytical. While holistic rubrics are generally used to obtain information about a general view of a student's performance during the term, analytic rubrics provide a separate evaluation of the products (for example, cell model, seed germination) that students have produced independently during the term. What is important here is that students should have prior knowledge of this assessment. It is clear that students who have knowledge of how to evaluate their own work will develop much more successful performances. Although rubrics are known as a very important assessment tool, the preparation and usage require a very long time to evaluate students' performances.

Concept Maps

Concept Maps (CM), from an assessment tool perspective, are two-dimensional diagrams that are used to measure important aspects of the structure of a student's declarative knowledge, assess conceptual meanings, identify pre- and alternative concepts. In the 1970s, Joseph Novak and a few Cornell University students created idea maps when promoting Ausubel's meaningful learning theory in a project (Novak & Gowin, 1984). Teaching abstract concepts is a common challenge for educators in the field of science education. The misconceptions surrounding 'abstract notions' are the most frequent issue arising during this process. Kinchin and Hay (2000) claim that structuring CMs is a useful metacognitive tool, which improves understanding and fosters opportunities to connect new knowledge to current structure in science education. For assessing the arrangement of information, CMs have also been employed in science and technology education (Rice, Ryan & Samson, 1998; Ruiz-Primo & Shavelson, 1996; White & Gunstone, 1992).

Mind Maps

According to Weideman and Kritzinger (2003), mind mapping is a method for representing and categorizing information that incorporates relationships, concepts and brain-friendly terms (Ehrlich, 2001). A mind map is composed of a main subject, such as a picture of the subject, and sub-branches containing keywords on related concepts, ideas and facts. When appropriate, the individual freely incorporates visual elements (such as images, forms and images) into their mind maps to help them to remember the concepts, thoughts and information contained therein (Buzan & Buzan, 1995; Proctor, 1999).

Concept Cartoons

Concept cartoons are designed to elicit students' ideas, challenge their thinking, and support the development of their understanding. They are drawings in the form of cartoons arranged as a stimulus to ask questions in the classroom. These tools present various perspectives on a topic within speech bubbles, including both legitimate scientific perspectives and widespread misconceptions, allowing students to identify and address them directly in science classes.

The Use of Web 2.0 Technologies in Assessment and Evaluation

Rapid developments in science and technology expand people's comfort zones in many areas. The new applications of science and technology in the field of education also provide greater convenience for teachers, students and school administrators. One of the discoveries from science and technology in education is Web 2.0 technologies. Web 2.0 refers to the second-generation web pages and applications that facilitate communication and provide secure information and collaboration on the Internet (Alexander, 2006). Web 2.0 technologies can be used for communication, interaction, information-sharing and easy access to information, collaborative content creation, content storage and sharing, evaluation and visualization (Sügümlü & Aslan, 2022). Web 2.0 technologies may enable participants to perform several different applications, such as challenging the existing status, responding to the questions and telling alternate tales (Buffington, 2008, p.307). Web 2.0 tools are currently going through a process in which they are widely used in the field of teaching and learning. Especially since March 2020, due to the effect of the pandemic and the shift of teaching and learning environments to distance education, it has been seen that both teachers and students have dramatically increased the use of technology in learning situations. In this process, it has been observed that teachers, especially, evaluate their students by using easy, effective and motivation-enhancing Web 2.0 tools. It has been reported in many studies that there have been various attempts by teachers to use Web 2.0 tools in their lessons (Alexander, 2006; Dalsgaard, 2006; Franklin & van Harmelen, 2007; Richardson, 2006).

In this section, information on how sample Web 2.0 tools can be used as an effective assessment tool in learning and teaching environments will be presented.

Kahoot, Quizizz, Socrative and Plickers are examples of Web 2.0 tools that can help and support teachers and educators to assess and evaluate learners' knowledge and skills effectively and efficiently within the context of classroom response systems.

The name of the Web 2.0 tool	Kahoot! https://kahoot.com/	
Explanation	Kahoot is a Web 2.0 tool that can be used in and out of classroom assessment. Due to its powerful features, it is one of the most popular Web 2.0 tools used for evaluation. All users must have a smart device to use this tool. Kahoot is a web platform powered by a user interface. Kahoot also has a mobile application.	

Properties	*Quick check:* Kahoot can help teachers and educators assess how the class feels about a topic and find out how students are really doing, which is absolutely essential when teachers teach remotely. *Assessing previous knowledge:* In constructivist learning theory, prior knowledge is very important in gaining new knowledge. Thanks to its feature, Kahoot can provide teachers with the level of knowledge of the students on the relevant subject through very simple tests. *More testing:* Teachers can share more tests with their students, helping students learn, especially about test techniques, and get more successful scores. *Misunderstandings:* Kahoot provides information to teachers about the misunderstandings of students and how to eliminate them. *Individual work:* Kahoot provides a better environment for students to study and practice at home or in class. *Enhanced analytics:* Kahoot has analytics that reports on and assesses learning outcomes and class progress.

The name of the Web 2.0 tool	Quizizz https://quizizz.com/	**QUIZIZZ**
Explanation	Quizizz is a powerful Web 2.0 tool similar to Kahoot. It is popularly used around the world, especially for formative assessment. In Quizizz, students are also required to use their own smart devices. Teachers, on the other hand, should prepare by working in advance on the questions that they will present to students on Quizizz. Tests can be administered in the classroom or answered at home as homework. In Quizizz, more reading processes can be recognized among students who have problems in solving tests in a short time. It is more important to answer more questions correctly than to answer in a short time to earn points, compared to Kahoot.	
Properties	*Question screen:* Students have the opportunity to see the questions on their own screen. Thus, students do not have any problems reading and answering the questions on any screen or presentation. *Location-independent:* There is the possibility of testing by ensuring that all students are on a single platform, regardless of location. *A large number of teacher-prepared tests:* There is access to tests prepared by teachers from many parts of the world. *Instant check:* The answers to the tests given to the students can be monitored instantly by the teacher.	

The name of the Web 2.0 tool	Socrative https://www.socrative.com/	socrative
Explanation	Socrative is a Web 2.0 application based on the logic of answering questions prepared by teachers via smart devices used by students. As a free application, Socrative reveals the extent to which students understand the subjects and offers detailed analyses to teachers. As in other applications in this field, instant answers given by students to questions in Socrative are automatically seen on smart devices used by teachers. The data obtained from the students can be used very easily for evaluation purposes.	
Properties	*Various events:* Socrative has a variety of events, including space races and exit tickets (a quick-check exercise that takes place in the last 5 min of a class). *Instant feedback:* Students' knowledge can be evaluated through activities previously prepared by teachers or instant questions. The information obtained from this provides teachers with ways to better plan their teaching. *Personalized activities:* Teachers can plan personalized activities and apply them to the whole class, as well as to individual students. *Saving time:* Student assessments are a seriously time-consuming task. By lessening this workload, Socrative helps teachers to find more time to plan their teaching better.	

The name of the Web 2.0 tool	Plickers https://plickers.com/	plickers plickers, simplified
Explanation	Plickers has different features from the Web 2.0 tools mentioned above. It is designed especially for students who do not have smart devices. Only the teacher's smart device, with an iOS or Android operating system, is sufficient to enable students to evaluate in the classroom. However, as in other assessment tools, there are aspects of this assessment tool that require the teacher to prepare beforehand. The teacher creates their questions online and prints Plickers cards for each student to use. The teacher poses questions in the classroom and asks the students to use their Plickers cards to show their answers. The teacher scans their smart device to find the correct answer to the question.	
Properties	*Plickers cards:* Thanks to Plickers' special cards, students do not need to have smart devices. This feature is an important technology that distinguishes Plickers from other Web 2.0 tools. *Formative/summative assessment:* Plickers can be defined as a formative and/or summative assessment tool that can help in assessing the cognitive and affective domains of students. *Data collection:* By collecting data from students easily, teachers can reveal cognitively which concepts students gain correctly and with which ones they have problems. *Learning process:* Plickers allow students to participate with high motivation and an interest in assessment activities prepared by the teacher and presented in the classroom.	

In addition to the tools mentioned above and used for evaluation purposes, applications and websites such as Edmodo, Nearpod, Quizlet, Microsoft Forms and QuizSocket are also important and can be used for assessment and evaluation purposes.

Undoubtedly, there are many advantages in using the above-mentioned assessment tools in the classroom. Many of the benefits have been outlined above. However, there are also many challenges in the use of these tools in and out of the classroom. These issues are briefly mentioned below:

- The use of Web 2.0 tools requires optimum use of smart devices. It has been reported in many studies that teachers with low IT literacy avoid using these applications (for example, Lena & Gurvitch, 2018). In this respect, it is very important for teachers to have the necessary in-service training for the use of Web 2.0 tools.
- In some schools, students' use of smart devices (for example, mobile phones) is limited or completely prohibited. In this case, it will not be possible to use all of the above tools (except Plickers). For this reason, various criteria should be introduced for the use of smart devices by students.
- It is important that teachers receive training to prepare quality questions. In such applications, questions that are incorrectly prepared, long, complex and repetitive will reduce the effectiveness of these Web 2.0 tools.
- Since most of these Web 2.0 tools will need to work in environments where wireless Internet is provided, Internet service provision in schools should be checked.
- In tests conducted against time, individual characteristics of some students should be taken into account. The fact that students with low individual response rates consistently receive low grades may negatively affect their interest and motivation in such assessment activities. In this respect, it is important to pay attention to the students' own response speeds.

Discussion and Recommendations

Assessment and evaluation play an important role in revealing the extent to which an educational process has achieved its purpose, determining the extent to which students have acquired the achievements, better planning learning and teaching environments, and ensuring the effective use of existing resources by the schools.

In determining these processes in a crystal-clear way, it is necessary to prepare the assessment tools that will provide the evaluation processes very seriously, to ensure their validity and reliability and to apply them appropriately.

Undoubtedly, the roles and responsibilities of teachers in these processes are vital. It is clear that a teacher who has in-depth knowledge in the field of assessment and evaluation can prepare quality assessment tools. In this context, teacher training institutions should take important responsibilities.

It is also necessary to increase the knowledge and skills of teachers, with in-service training on new and alternative assessment tools. In this regard, planning and implementation of the necessary in-service teacher training by the ministries of national education will make an important contribution to the formation of quality education processes in the countries.

Increasing the use of new technologies (such as Web 2.0 tools) as assessment tools in learning and teaching environments will provide a positive classroom atmosphere, instant assessment data of students' knowledge and skills, and quality learning and teaching environments. This will also provide quality learning gains for students.

Summary

The following ideas have been discussed in this chapter: (1) featured explanations of assessment and evaluation in science and technology education; (2) suggestions of practical assessment tools and examples to be used in science and technology education; (3) support for teacher assessment and evaluation literacy; (4) suggestions of Web 2.0 tools that can be used to assess students' gains in science and technology; and (5) discussions and recommendations of related books and journal articles in science and technology education (see below).

Disclosure

The authors are not affiliated in any way with Kahoot, Quizziz, Socrative and Plickers, and there is no conflict of interest.

Recommended Resources – Books and Journal Articles

The authors recommend the following books and journal articles that provide further information on assessment and evaluation in science and technology education:

Books:
Chittenden, E. (1991). Authentic assessment, evaluation, and documentation of student performance. In Perrone, V. (Ed.), *Expanding student assessment* (pp. 22–31). Association for Supervision and Curriculum Development.
Dolin, J. & Evans, R. (Eds.) (2017). *Transforming assessment: Through an interplay between practice, research and policy (Vol. 4).* Springer.
Earle, S. (2019). *Assessment in the primary classroom: Principles and practice.* Learning Matters.

Enger, S. K. & Yager, R. E. (2009). *Assessing student understanding in science: A standards-based K-12 handbook.* Corwin Press

Liu, X. (2010). *Essentials of science classroom assessment.* Sage Publications. https://doi.org/10.4135/9781483349442

Journal Articles:

Erduran, S., El Masri, Y., Cullinane, A. & Ng, Y. P. D. (2020). Assessment of practical science in high stakes examinations: a qualitative analysis of high performing English-speaking countries. *International Journal of Science Education, 42*(9), 1544–1567.

Klosterman, M. L. & Sadler, T. (2010). Multi-level Assessment of Scientific Content Knowledge Gains Associated with Socioscientific Issues-based Instruction. *International Journal of Science Education, 32*(8), 1017–143.

Zhai, X., He, P. & Krajcik, J. (2022). Applying machine learning to automatically assess scientific models. *Journal of Research in Science Teaching,* 1–30. https://doi.org/10.1002/tea.21773

References

Alexander, B. (2006). Web 2.0: A new wave of innovation for teaching and learning? *Educause Review, 41*(2), 32–44.

Boonchutima, S., & Pinyopornpanich, B. (2013). Evaluation of public health communication performance by Stufflebeam's CIPP model: A case study of Thailand's department of disease control. *Journal of Business and Behavioral Sciences, 25*(1), Article 36.

Buffington, M. L. (2008). Creating and consuming Web 2.0 in art education. *Computers in the Schools, 25*(3–4), 303–313.

Buzan, T., & Buzan, B. (1995). *The mind map book.* BBC Books.

Çepni, S., & Ayvacı, H. Ş. (2016). Measurement and evaluation in science and technology education. In S. Çepni (Ed.), *Teaching science and technology from theory to practice.* Pegem Publishing.

Dalsgaard, C. (2006). Social software: E-learning beyond learning management systems. *European Journal of Open, Distance and E-Learning,* (2). Retrieved from: https://www.assonur.org/sito/files/Social%20Software%20as%20learning%20tool.pdf. Accessed 25 July 22.

Ehrlich, A. R. (2001). *Mind mapping: An overview bibliography.* Retrieved from: https://www.academia.edu/387624/The_use_of_mind_mapping_technique_in_chemistry_teaching. Accessed 25 July 22.

Franklin, T. & van Harmelen, M. (2007). Web 2.0 for content for learning and teaching in higher education. *JISC Report* [verified 21.02.10]. http://www.jisc.ac.uk/media/documents/programmes/digitalrepositories/Web2-contentlearning-and-teaching.pdf

Gipps, C., & Stobart, G. (2003). Alternative assessment. In T. Kellaghan & D. L. Stufflebeam (Eds.), *International handbook of educational evaluation* (pp. 549–575). Springer.

Kinchin, I. M., & Hay, D. B. (2000). How a qualitative approach to concept map analysis can be used to aid learning by illustrating. *Educational Research, 42*(1), 43–57.

Kreisman, S., Knoll, M., & Melchior, T. (1995). Toward more authentic assessment. In A. L. Costa & B. Kallick (Eds.), *Assessment in the learning organization* (pp. 114–138). Association for Supervision and Curriculum Development.

Lena, C., & Gurvitch, R. (2018). Using plickers as an assessment tool in health and physical education settings. *Journal of Physical Education, Recreation & Dance, 89*(2), 19–25.

Martin, J., & Collins, R. (2011). Formative and summative evaluation in the assessment of adult learning. In V. C. X. Wang (Ed.), *Assessing and evaluating adult learning in career and technical education* (pp. 127–142). IGI Global. https://doi.org/10.4018/978-1-61520-745-9

McTighe, J. & Ferrara, S. (1998). *Assessing learning in the classroom. Student Assessment Series*. NEA Professional Library, Distribution Center, PO Box 2035, Annapolis Junction, MD 20701-2035

MoNE. (2017). *Science course curriculum (For primary and secondary schools 3, 4, 5, 6, 7, 8th grades)*. Ministry of National Education.

National Research Council. (2013). *Next generation science standards: For states, by states*. The National Academies Press. https://doi.org/10.17226/18290

Novak, J. D., & Gowin, R. (1984). *Learning how to learn*. Cambridge University.

Proctor, T. (1999). *Creative problem-solving for managers*. Routledge.

Richardson, W. (2006). *Blogs, wikis, podcasts, and other powerful web tools for classrooms*. Corwin Press.

Rice, D. C., Ryan, J. M., & Samson, S. M. (1998). Using concept maps to assess student learning in the science classroom: Must different methods compete? *Journal of Research in Science Teaching, 35*(10), 1103–1127.

Ruiz-Primo, M. A., & Shavelson, R. (1996). Problems and issues in the use of concept maps in science assessment. *Journal of Research in Science Teaching, 33*, 569–600.

Shanklin, N., & Conrad, L. (1991). *Portfolios: A new way to access student growth*. Colorado Council of the International Reading Association.

Stiggins, R. J. (1994). *Student-centered classroom assessment*. Macmillan.

Sügümlü, Ü., & Aslan, S. (2022). The use of Web 2.0 tools in mother-tongue instruction: Teachers' experiences. *International Journal of Education and Literacy Studies, 10*(1), 124–137.

Weideman, M., & Kritzinger, W. (2003). *Concept mapping – A proposed theoretical model for implementation as a knowledge repository*. ICT in Higher Education.

White, R. T., & Gunstone, R. F. (1992). *Probing understanding*. Falmer Press.

Yambi, T. (2018). *Assessment and evaluation in education*. https://www.academia.edu/35685843/

Professor Dr. Bulent Çavaş completed his Master and PhD studies in the field of science education at Dokuz Eylul University, Faculty of Education, Science Teacher Training Programme in 1998. He did his post-Doc studies in Middle East Technical University. He has produced over 150 national and international publications and written 10 books on science and science education. He is Secretary of the CMAS-Science Committee. Currently, his research interests are Responsible Research and Innovation, Open Schooling, Inquiry-Based Science Education and Virtual Reality in Science Education. He is one of the past Presidents of the International Council of Associations for Science Education (ICASE – www.icaseonline.net). He works as an external expert for evaluating European Commission projects in Brussels, Belgium. Currently, he is working as Professor of Science Education at Dokuz Eylul University (www.deu.edu.tr) in Izmir, Turkey.

Professor Dr. Pınar Çavaş has a Bachelor's degree from Ege University, Physics Department. She continued her Master's at Ege University on primary education (major: science education). She completed her PhD degree at Dokuz Eylul University, working on the elementary teachers' scientific literacy level. She has more than 100 journal articles, conference papers, books and book chapters. She was involved in some European Union projects, as well as university-supported projects. Her fields of interest are primary science education, scientific literacy, and teachers' competences. In addition to her director role at the Children's Education Application and Research Center, she is heading the Department of Primary Education, Faculty of Education, Ege University (www.ege.edu.tr).

Professor Dr. Şengül Anagün completed her Master's degree in the field of educational administration. She obtained her PhD in the field of Primary Education at Anadolu University-Turkey. Her research interests are science teaching in primary education, curriculum and instruction, and teacher training. She has published many articles and book chapters on primary science education. She is an editor of three science education books. She has experience as a teacher in primary education and has been teaching at the Faculty of Education for more than 25 years. She works as Professor in the Primary Education Department at Eskisehir Osmangazi University in Eskisehir, Turkey. Currently, she is working as a Visiting Professor at Ege University in Izmir, Turkey.

Chapter 7
Mathematics in the Service of Science and Technology Education

Ajeevsing Bholoa and Ajay Ramful

Abstract STEM subjects are regarded as pivotal for the transformation of modern societies, for the enhancement of the quality of life, for addressing challenges jeopardizing human existence, for economic survival and for ensuring global security. Recognizing that mathematics is the bedrock of science and technology, there has been noticeable investment in the teaching and learning of mathematics. However, we still face the dual challenge of addressing the relatively mild engagement of students in school mathematics and preparing them for STEM subjects. The current chapter puts into perspective this dual challenge and situates the prospects and possibilities that twenty-first century mathematics offer as a service subject to science and technology. We reinforce the call for re-engineering school mathematics, to move beyond the traditional conception of mathematics as being a subject of rules and procedures to one that offers the knowledge and skills to solve contemporary problems, create and innovate in the service of science and technology. We conclude with some teaching and learning proposals for school mathematics in its auxiliary role for STEM subjects.

Keywords Twenty-first century mathematics · School mathematics curriculum · Science and technology · STEM education

Introduction and Chapter Map

Although they are intricately related, science and technology are distinct disciplines with particular affordances (Pleasants et al., 2019). While science explores new knowledge, technology opens new avenues for the application of scientific knowledge. Throughout human history, mathematics has played a vital and critical role in

A. Bholoa (✉) · A. Ramful
Mauritius Institute of Education, Moka, Mauritius
e-mail: a.bholoa@mie.ac.mu; a.ramful@mie.ac.mu

© The Author(s), under exclusive license to Springer Nature Switzerland AG 2023
B. Akpan et al. (eds.), *Contemporary Issues in Science and Technology Education*, Contemporary Trends and Issues in Science Education 56, https://doi.org/10.1007/978-3-031-24259-5_7

spurring on and supporting innovations in science, technology and everyday life. Historical examples of the contribution of mathematics in scientific and technological innovations include the invention of zero, logarithm, complex numbers, calculus or Euclidean geometry, which have allowed the precise quantification and understanding of natural phenomena. Mathematical sciences that essentially consist of mathematics, statistics, operations research and theoretical computer science are widely used to reflect and represent the multiple applications of mathematics in modern communication, transportation, science, engineering, technology, medicine, manufacturing, security and finance (National Research Council, 2012). The most recent 'needs-based' example (Anderssen et al., 2016) of mathematics in science and technology could be seen during the outbreak of the COVID-19 pandemic, involving simulation models of the transmission dynamics and spread of coronavirus. Alternatively, an 'idea-based' model of mathematics, as robotics engineering, can be utilized to design models that experiment with recent developments in the field of science and technology (Anderssen et al., 2016).

At the school level, efforts in the application of mathematics in practical and scientific problems and technology education are reflected in Applied Mathematics. The problems are often contextualized as word problems requiring factual, conceptual and procedural understanding of a static or immutable body of mathematical knowledge, which has been passed on from one generation to the next (Dooley & Corcoran, 2007). However, it appears that the response to integrating twenty-first century mathematics into the school curriculum has been relatively slow.

This chapter brings into focus the pivotal role of twenty-first century mathematics for the enhancement of science and technology education. Firstly, it elaborates the key components of twenty-first century mathematics. This new set of knowledge is examined in relation to secondary school mathematics to show the possibilities and challenges for its integration into the curriculum. The school mathematics curriculum is discussed, with a view to establishing its responsiveness to scientific and technological innovations.

Furthermore, comparisons of the GCSE mathematics (aged 16+ years) and A-level mathematics (aged 18+ years) curriculum, spanning over a period of more than 20 years, are correspondingly made to show the degree of responsiveness of the curriculum. It is observed that, over the two decades, the emphasis of the mathematics curriculum has been on foundational knowledge with minimal attention to twenty-first century mathematical concepts. In the concluding section, some pointers are provided to infuse twenty-first century mathematics to better serve science and technology education.

Early Contributions of Mathematics to Science and Technology

Since antiquity, mathematics has been fundamental to advances in science and engineering. From simple counting, calculation and measurement, mathematics has evolved to encompass a range of specialized areas related to science and technology (National Academies of Sciences, Engineering and Medicine, 1991). Table 7.1 illustrates selected key mathematical developments from the Babylonian era

Table 7.1 History of development of mathematical techniques

	Sample mathematical techniques	Applications to science and technology	References
Babylonian mathematics	Basic arithmetic Co-ordinate system Base 60 number system	Motion of planets Clock	Blaken (1986)
Egyptian mathematics	Linear measurement Multiplication by binary factors	Scaled rods Development of computer Building of pyramid	Clagett (1999)
Greek mathematics	Pythagoras' theorem Archimedes' *The Method*	Study of astronomy Mechanical experiments	Violatti (2013)
Indian mathematics	Trigonometry Concept of zero Rules of negative numbers Differential calculus	Land surveying Navigation Application to astronomy problems	Yates (2017)
Seventeenth century mathematics	Logarithm Analytic geometry Probability Infinitesimal calculus Power series	Logarithm slide rule Astronomy Laws of physics Optics Computer	Knorr et al. (2020)
Eighteenth century mathematics	Probability Complex numbers Descriptive geometry Differential equations	Classical mechanics Mechanical drawing Engineering Particle dynamics Theory of fluids	Knorr et al. (2020)
Nineteenth century mathematics	Fourier series Non-Euclidean geometry Boolean algebra Chaos theory Probability theory Least squares Linear algebra	Theoretical physics Mechanics Cartography Computer science Quantum electrodynamics Electromagnetism	Knorr et al. (2020)
Twentieth century mathematics	Set theory Maxwell's equations Integral equations Vector spaces Probability theory Random graphs	Quantum mechanics Theory of relativity Finance and trading Thermodynamics Statistical mechanics Computer science	Knorr et al. (2020)

(3000 BC – 260 AD) to twentieth century mathematics and its application to science and technology.

Table 7.1 illustrates the evolving and dynamic nature of mathematics as a body of knowledge connected to scientific and technological advancement. These developments have been accentuated in an unprecedented way in the twenty-first century, primarily marked by the fourth industrial revolution.

Twenty-First Century Mathematics in Service to Science and Technology

Twenty-first century mathematics has largely been built upon the achievements of the past centuries, but increasingly with sophisticated applications to science and technology. The possibilities that twenty-first century mathematics offer for the modelling and resolution of problems are gigantic (see Fig. 7.1), especially with the affordance of advanced computing facilities and new areas such as Artificial Intelligence (AI).

The twenty-first century has also witnessed the development of the fourth industrial revolution and the contribution of mathematics to embrace this revolution has been intensively studied, ranging from the professional development of teachers to the mathematics curriculum in schools or universities. For instance, in their study, Ramful and Patahuddin (2021) compared the contemporary school mathematics curriculum with a projected one that would not only lay the mathematical foundation for learners to operate the technologies, but also ensure that they have *'a problem-solving attitude beyond mastering concepts and procedures'* (p.16). Although the foundational knowledge derived from the contemporary school mathematics curriculum (numbers, geometry, measurement, algebra, probability, statistics, calculus…) cannot be undermined, there is a strong call for integrating mathematical knowledge and skills necessary to handle the fourth industrial revolution (Formaggia, 2017) (see Fig. 7.2).

More specifically, Gravemeijer et al. (2017) indicate potential mathematical content required to raise the need for skills to complement science and technology education. This includes:

- statistics (big data, statistical literacy, data collection, variables, variation);
- space geometry (3D imaging, 3D printing, spatial reasoning);
- mathematical models (functions, numerical analysis, programming);
- numeracy and quantitative literacy (pattern, function, variability, sampling, probability, prediction, data displays); and
- number theory (coding, hacking).

Along a similar vein, a recent study conducted to investigate the required changes in Saudi universities' mathematics curricula to satisfy the country's Vision 2030 (Alabdulaziz, 2019) reported that *'algebra may be applied to computer sciences,*

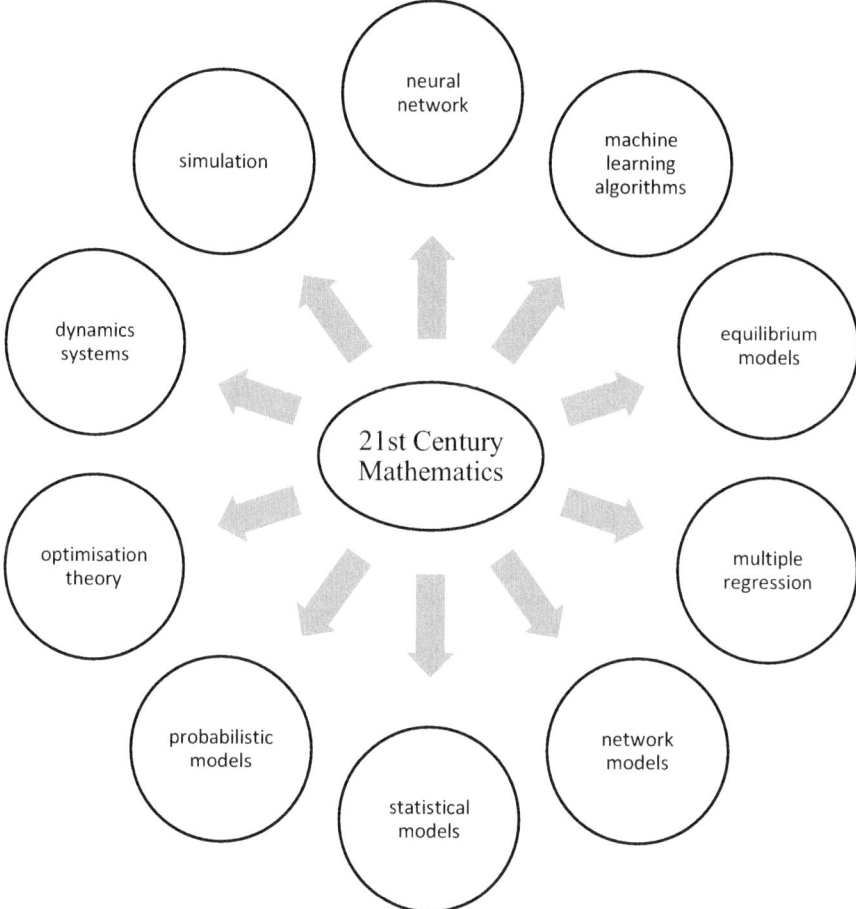

Fig. 7.1 Twenty-first century mathematical techniques

networking, cryptology, and study of symmetry in physics and chemistry. Calculus (differential equations) may be applied to biology, engineering, physics, molecular structure, rocket science, the motion of water as well as option price modelling in economics and business paradigms'.

UNESCO, in its recent publication entitled *Mathematics for action: supporting science-based decision-making* (2022), recognised the power of mathematics in the twenty-first century for ensuring sustainable development. It principally describes and connects the contribution of mathematics in terms of the goals of the 2030 Agenda for Sustainable Development (see Chap. 10). In Table 7.2, we analyse UNESCO's toolkit to illustrate the relationship between the emerging twenty-first century mathematical techniques and topics with science and technology. It is evident that science and technology education requires mathematical knowledge and understanding beyond that which school mathematics currently offers. Concepts

Fig. 7.2 Twenty-first century mathematical knowledge and content

related to linear algebra, real analysis, differential equations, probability and statistics are recurrent in contemporary applications and demand relatively higher-order mathematical knowledge and skills in the school mathematics curriculum.

Although the discipline of mathematics has evolved rapidly over time (as shown in Table 7.1), the school mathematics curriculum could not keep pace with such changes, widening the gap between contemporary demands of science and technology and mathematical readiness (Bholoa & Ramma, 2019). In the next section, we compare the school mathematics curriculum for GCSE and A-level curricula at two-decades' intervals.

Table 7.2 2030 agenda connecting mathematics, science and technology

Goal	Description	Connection to science and technology education	Twenty-First century mathematical techniques	Mathematical topics
1	End poverty in all its forms everywhere	Visualizing poverty using artificial intelligence-powered maps to improve estimates and predictions Estimates are merged with a 'poverty' map using a geographic information system (GIS)	Machine learning algorithms Artificial intelligence Multiple regression Spatial relationships	Linear algebra Probability Statistics Calculus Discrete maths
2	End hunger, achieve food security and improved nutrition and promote sustainable agriculture	Strengthening food security to build sustainable, productive and resilient food systems Simulating antibiotic resistance control policy in agriculture	General equilibrium models Network models	Logic and set theory Euclidean space Real analysis
3	Ensure healthy lives and promote wellbeing for all at all ages	Forecasting the likely impact of an epidemic Estimate a measure of the contagiousness of a pathogen Predicting the effectiveness of measures of disease containment and prevention Artificial intelligence technologies and tools can play a key role in pandemic response and public health decision-making Enhancing the design of effective new vaccines and enabling the redesign of existing ones Building phylogenetic trees that illustrate how the genetic sequences of circulating viruses are related to that of the current vaccine virus	Epidemiological models Artificial intelligence Statistical models Data assimilation methods Game-theoretical models	Linear algebra Ordinary differential equations Systems of differential equations Probability theory Stochastic processes (theory of random variables) Numerical methods

(continued)

Table 7.2 (continued)

Goal	Description	Connection to science and technology education	Twenty-First century mathematical techniques	Mathematical topics
6	Ensure availability and sustainable management of water and sanitation for all	Quantifying risks and identifying appropriate options for the management of water supply and quality Supporting successful and sustained management of important ecosystems by providing fundamental insights into their dynamics and vulnerabilities	Bayes' theorem	Probability Bayesian statistics
9	Build resilient infrastructure, promote inclusive and sustainable industrialization and foster innovation	Improving resilience of digitalised complex systems Modelling, simulating, and assessing the behaviour of critical infrastructure components	Probabilistic model (e.g., CASCADE model) Network optimisation	Probability Programming Optimisation problems
12	Ensure sustainable consumption and production patterns	Strengthening assessments and projecting future changes in ecosystem services Providing policy-relevant information to ensure benefits to future generations	Dynamic systems (e.g., predator-prey difference equations) Stability theory	Optimisation Theory of differential equations Theory of dynamical systems
16	Promote peaceful and inclusive societies for sustainable development, provide access to justice for all and build effective, accountable and inclusive institutions at all levels	Meeting the requirements of data protection regulations Enhancing disease detection tools Identifying and deterring financial fraud	Machine learning (neural networks)	Linear algebra Geometry Calculus Optimization Probability and statistics

School Mathematics Curriculum

Mathematics has been a fundamental part of the school curriculum. Dossey (1992) states that '*Perceptions of the nature and role of mathematics held by our society have a major influence on the development of school mathematics curriculum, instruction, and research*'. However, mathematics has been consistently viewed as a static subject (Fisher, 1990) and watered down (Siddiqi et al., 2011) over the years, despite the fact that the need for sophisticated mathematical understanding and modelling has become more pressing due to the increasingly complex nature of scientific and technological applications. As reported by the National Research Council (2013), the education system may not successfully prepare students for college, careers and citizenship. It called for a review of the learning expectations and goals so that learners, right from kindergarten, develop an understanding of physical sciences, life sciences, space sciences, engineering, technology and application of science. However, education systems have largely been slow to respond to modern challenges (Odell & Pedersen, 2020).

As an illustration of the evolution of mathematics as a school subject, Table 7.3 shows the comparison of the topics/themes in the GCSE syllabus for school mathematics offered by the University of Cambridge Local Examinations Syndicate in

Table 7.3 Comparison of GCSE mathematics syllabuses of 2000 and 2022

Topic/Theme in GCSE syllabus 2000	Topic/Theme in GCSE syllabus 2022
Numbers	Number
	Set language and notation
Squares, square roots, cubes and cube roots	Squares, square roots, cubes and cube roots
	Directed numbers
Vulgar and decimal fractions and percentages	Vulgar and decimal fractions and percentages
Ordering	Ordering
Standard form	Standard form
The four operations	The four operations
Estimation	Estimation
	Limits of accuracy
Ratio, proportion, rate	Ratio, proportion, rate
Percentages	Percentages
Use of scientific calculator	Use of scientific calculator
	Time
	Money
	Personal and small business finance
Everyday mathematics	
Graphs in practical situations	Graphs in practical situations
Graphs of functions	Graphs of functions
	Function notation
Co-ordinate geometry	Co-ordinate geometry
Algebraic representations and formulae	Algebraic representations and formulae
Algebraic manipulation	Algebraic manipulation
Indices	Indices
Solutions of equations and inequalities	Solutions of equations and inequalities
	Graphical representations of inequalities
	Sequences
	Variation

(continued)

Table 7.3 (continued)

Topic/Theme in GCSE syllabus 2000	Topic/Theme in GCSE syllabus 2022
Geometrical terms and relationships	Geometrical terms
Geometrical constructions	Geometrical constructions
	Similarity and congruence
Bearings	
Symmetry	Symmetry
Angle	Angles
Locus	Loci
	Measures
Mensuration	Mensuration
Trigonometry	Trigonometry
Statistics	
	Categorical, numerical and grouped data
	Statistical diagrams
Probability	Probability
Transformations	Transformations
Vectors in two dimensions	Vectors in two dimensions
	Matrices

Box 7.1 The GCSE Mathematics Syllabus Aims in 2000 (Syllabus 4107)

- develop mathematics language as a means of communication
- acquire a foundation appropriate to a further study of mathematics and skills and knowledge pertinent to other disciplines
- acquire and apply skills and knowledge relating to number, measure and space in mathematical situations that they will meet in life
- develop an understanding of mathematical principles and the abilities to reason logically
- conduct individual and co-operative enquiry and experiment, including extended pieces of work of a practical and investigative kind
- integrate information technology to enhance the mathematical experience
- engage in imaginative and creative work arising from mathematical ideas
- enhance intellectual curiosity and appreciate the power and structure of mathematics, including patterns and relationships
- develop a positive attitude towards mathematics, including confidence, enjoyment and perseverance
- appreciate the interdependence between the different branches of mathematics

---

2000 and 2022. It can be observed that most of the topics/themes have remained unchanged or re-branded over the span of more than 20 years. However, more importantly, it is the syllabus aims of these two periods that have witnessed significant changes (see Boxes 7.1 and 7.2).

Box 7.2 The GCSE Mathematics Syllabus Aims in 2022 (Syllabus 4024)

- increase intellectual curiosity, develop mathematical language as a means of communication and investigation and explore mathematical ways of reasoning
- acquire and apply skills and knowledge relating to number, measure and space in mathematical situations that they will meet in life
- acquire a foundation appropriate to their further study of mathematics and of other disciplines
- appreciate the pattern, structure and power of mathematics and derive satisfaction, enjoyment and confidence from the understanding of concepts and the mastery of skills

However, in comparison to 2000, the aims of the syllabus in 2022 (4024) (see Box 7.2) have laid lesser importance on some components related to science and technology education and the twenty-first century skills of learners. For example, less focus is given to conducting individual and co-operative mathematical inquiry and experiment, enhancing and exploring mathematical experiences using information technology and creative mathematical work.

Besides the slow evolution of school mathematics over time, there is also evidence of the watering down of the school mathematics curriculum, as can be observed from the comparison of topics in the A-level syllabus for the years 2000 and 2022. For example, for the topic 'Numerical solution and equations', which is a fundamental component of mathematical modelling in science and technology education, about two decades ago students were required to understand, in geometric terms, the working of the Newton-Raphson's method and derive and use iterations based on that method. In 2022, students are merely required to understand how a given simple iterative formula relates to the equation being solved, and use a given iteration, or an iteration based on a given rearrangement of an equation. Similarly, compared to 2000, the A-level syllabus in 2022 does not make provision, amongst others, for:

- Maclaurin series expansions;
- Knowledge of function (periodicity and symmetries) of inverse trigonometric functions;
- Concepts related to bivariate data, such as regression lines, correlation coefficient; and
- statistical interference such as t-test, chi-square test for independence.

A summary of important changes to the A-level syllabus, which illustrates the 'static' and the 'watering-down' perspectives, is shown in Table 7.4.

Table 7.4 Comparison of A-level mathematics syllabus of 2000 and 2022

Topic (P1)	Year 2000	Year 2022
Quadratics	All concepts remained the same	All concepts remained the same
Functions	All concepts remained the same	All concepts remained the same
Co-ordinate geometry	Linear law	Equations of circles Use algebraic methods to solve problems involving lines and circles Including use of elementary geometrical properties of circles, e.g., tangent perpendicular to radius, angle in a semicircle, symmetry
Circular measure	All concepts remained the same	All concepts remained the same
Trigonometry	Periodicity and symmetries Use the concepts of period and/or symmetries in relation to these functions and their inverses	Not in syllabus
Series	Work with McLaurin series of simple functions such as $e^x sinx$, $ln(3 + 2x)$ (derivation of a general term is not included)	Not in syllabus
Differentiation	Understand the gradient of a curve at a point as the limit of the gradients of a suitable sequence of chords	Use information about stationary points in sketching graphs
Integration	All concepts remained the same	All concepts remained the same
Algebra	Indices and proportionality	Partial fractions
Logarithmic and exponential functions	All concepts remained the same	All concepts remained the same
Trigonometry	All concepts remained the same	All concepts remained the same
Differentiation	Not in syllabus	Use derivatives of $tan^{-1}x$
Integration	All concepts remained the same	All concepts remained the same
Numerical solution of equations	Trapezium rule	Not in syllabus
Vectors	Not in syllabus	Determine whether two lines are parallel, skewed or intersecting
Differential equations	Curve sketching	Not in syllabus
Complex numbers	All concepts remained the same	All concepts remained the same
The Poisson distribution	Not in syllabus	Use the normal distribution with continuity correction as an approximation to the Poisson distribution where appropriate
Linear combinations of random variables	All concepts remained the same	All concepts remained the same

(continued)

Table 7.4 (continued)

Topic (P1)	Year 2000	Year 2022
Continuous random variables	All concepts remained the same	All concepts remained the same
Hypothesis tests	t-tests	Type I, type II errors
Bivariate data	Least squares Regression lines Correlation	Not in syllabus
χ^2 tests	χ^2 tests Goodness of fit	Not in syllabus

Rethinking the Twenty-First Century Mathematics School Curriculum

Traditionally, school mathematics has revolved around five main strands – numbers, geometry, algebra, measurement, handling data (statistics) – based on the spiral curriculum, which gradually deepens the knowledge of mathematics concepts to elevate students to higher levels of abstraction (Fried & Amit, 2005). Usiskin (2007) argued that traditional word problems, geometric proofs and some algebraic manipulation should be replaced to give way for problems situated in realistic and contemporary settings, and the use of a computer algebra system (CAS) and geometry software for exploration purposes to facilitate teaching and learning of concepts in science and technology.

The National Council of Teachers of Mathematics (2000) highlighted that school mathematics remained disconnected from real life and from other subjects in the school curriculum, thus affecting students' ability to apply mathematical ideas to topics in the teaching of science and technology. The National Research Council asserts that a re-conceptualisation of the mathematics curriculum is needed, which prepares students for a rapidly-changing future that renders the current knowledge and skills in the workforce obsolete (National Research Council, 2012). The job market in the modern era is characterised by scientific and technological innovations largely driven by STEM fields (Khalil & Osman, 2017). Accordingly, much emphasis is being laid on the prioritisation, development and integration of STEM education in the school curriculum (see Chap. 4) as a response to contemporary needs (Fomunyam, 2020).

A similar view is echoed by Gravemeijer et al. (2017), who argue that with the increased availability of technology, routine skills (such as arithmetical computation and algebraic solution) will diminish. Furthermore, considering the growing fields such as Artificial Intelligence, Space Exploration, Robotics, and Material Science among others, Gravemeijer et al. (2017) contend that the future of mathematics education must be considered within the context of science, technology, engineering and mathematics (commonly referred to as STEM) and further provide the necessary mathematical competencies required, as displayed in Table 7.5.

Table 7.5 Mathematical competencies for STEM education

Category	Mathematical competencies
Applying/ modelling	Decoding and interpreting information, structuring and conceptualizing the problem situation, making inferences and assumptions, and formulating a model Generating a model to interpret, explain, and make predictions about that situation in order to solve the problem
Understanding	Working on the conceptual mathematical understanding that is needed to grasp the mathematics hidden in the digital tools that are abundant in our technological society
Checking	Checking mathematical correctness (adequacy of the mathematical procedures, approximations, or generalizations, validity and credibility of the results of statistical procedures…)

The above suggestions point to the fact that the key to integrating twenty-first century mathematics in the curriculum is through the infusion of thinking processes congruent to the underlying mathematical principles. For instance, computational thinking might be introduced as an activity supplementing the topic 'numerical solutions of equations'. It should also be acknowledged that the curriculum space may be quite constraining to further load the mathematical syllabus content supporting science and technology education. However, the parallel learning experienced could be promoted through STEM education.

Promoting STEM Education

In the literature, varied definitions of STEM education have been proposed by different researchers. One such definition is as follows (Fomunyam, 2020):

> *STEM education is the purposeful integration of STEM disciplines with the objectives of expanding students' abilities by supporting technical and scientific education* (p. XII).

Through careful design of the mathematics curriculum within the context of STEM, the teaching of mathematics serves primarily as a tool for the acquisition of STEM literacy (National Governors Association, 2006), such as:

- Scientific literacy: the ability to use scientific knowledge (in physics, chemistry, biology and space sciences) and processes to understand the natural world and phenomena and participate in decision-making;
- Technological literacy: the ability to use, manage, understand and assess technology and analyze how new technologies affect the world around us; and
- Engineering literacy: the ability to understand how technologies are developed via the engineering design process.

Project-based learning (PBL) can provide an effective model for the successful integration of STEM education in the school curriculum (Odell & Pedersen, 2020).

Project-Based Learning

Teachers can incorporate PBL in the learning experience through the use of ill-defined problems – problems that simply do not have enough information to be solved – and which can lead students to enhance their conceptual understanding and become more literate in subject areas (Ronis, 2008; Trilling & Fadel, 2009). PBL helps to develop logical and critical thinking skills of learners such as: Comparing and contrasting, Classifying, Sequencing, Drawing inference, Predicting outcomes, Identifying and creating patterns and symmetry and Drawing and interpreting charts and graphs (Ronis, 2008). The above-mentioned skills are strongly linked with the curricular goals and expectations of teaching and learning of mathematics. We provide three examples of how PBL can be used to support the curricular goals of contemporary science and technology education while consolidating twenty-first century mathematics:

- **Robotics**
  Robotics can bridge the four fields of STEM and is already being applied in schooling systems to enact mathematical concepts and support understanding of scientific and engineering principles in action (Leoste & Heidmets, 2019). Mathematical areas, such as algebra, geometry and calculus (graphs of functions) can be meaningfully applied to science concepts such as kinematics (distance, velocity, acceleration), force, torque, moment, energy, and understanding of technological concepts such as electric motors and gearboxes.
- **3D printing**
  Mathematical concepts related to 3D printing are closely associated with geometry, including surface area, volume of revolution, Cartesian planes and vectors, and provide a strong foundation for the understanding of physical models and technologies such as 3D scanners and computer-aided design (CAD). Science projects have also been proposed to explore concepts such as pendulum (simple harmonic motion), models of geological formations, moment of inertia and gravitational waves, among others (Hovarth & Cameron, 2017).
- **Material science**
  Projects in material science integrate various areas of chemistry and physics, including topics such as metals, polymers, ceramics, solid state physics, semiconductors and sustainable energy. Mathematical ideas and methods can be applied to materials problems – for example, computational homology applied to structural analysis of glassy materials, stochastic models for the formation process of materials, new geometric measures for finite carbon nanotube molecules, mathematical techniques predicting a molecular magnet, and network analysis of nonporous materials (Ikeda & Kotani, 2015).

Summary

In this chapter we affirmed that, while Mathematics has been fundamental in advancing scientific and technological innovations since antiquity through evolving mathematical techniques, the school mathematics curriculum has not been very responsive to these changes. School mathematics is still being viewed as static and considered to be watered down, while the requirements for science and technology are increasingly sophisticated. The integration of STEM education in the school curriculum facilitates the service of mathematics in science and technology education and project-based learning provides an effective model to guide the integration of STEM education at the school level. To meet the requirements of the 2030 Agenda for Sustainable Development, UNESCO has mapped the useful mathematical techniques that reflect the current and future needs relevant to science and technology education. These mathematical techniques are strongly associated with teaching and learning and understanding of science and technology concepts such as machine learning algorithms, neural network, multiple regression, general equilibrium models, network models, statistical models, probabilistic models, theory of optimisation, dynamic systems and simulation (see Fig. 7.1). Furthermore, the mathematical knowledge and content necessary for the study of science and technology include mathematical modelling, computing and programming, game theory, number theory, cryptography, numerical analysis, differential equations, operations research and optimisation, linear algebra, matrix theory, networks, inferential statistics, Bayesian statistics, probability theory, simulation and logic (see Fig. 7.2).

Recommended Resources

Bholoa, A., & Ramma, Y. (2019). Mathematics and science education. In B. Akpan (Ed.), *Science education: Visions for the future* (pp. 117–131). Next Generation Education.

Gravemeijer, K., Stephan, M., Julie, C., Lin, F. L., & Minoru, O. (2017). What mathematics education may prepare students for the society of the future?. *International Journal of Science and Mathematics Education, 15*(2), 105–123.

National Academies of Sciences, Engineering and Medicine. (1991). *Mathematical sciences, technology and economic competitiveness.* The National Academies Press.

National Council of Teachers of Mathematics. (2000). *Principles and standards for school mathematics.* NCTM.

National Governors Association. (2006). *Building a science, technology, engineering and math agenda.* Innovation America.

National Research Council. (2012). *Fuelling innovation and discovery: The mathematical sciences in the twenty-first century.* The National Academies Press.

National Research Council. (2013). *Next generation science standard: For states, by states.* The National Academies Press.

Siddiqi, A. H., Singh, R. C., & Manchanda, P. (2011). Mathematics in science and technology: Mathematical methods, models and algorithms in science and technology. In *Proceedings of the satellite conference of International Congress of Mathematics (ICM) 2010. International Congress of Mathematics.* World Scientific Publishing.

UNESCO. (2022). *Mathematics for action: Supporting science-based decision-making.* UNESCO.

Usiskin, Z. (2007). What should not be in algebra and geometry curricula of average college-bound students?": A retrospective after a quarter century. *Mathematics Teacher,* (100), 78–79.

References

Alabdulaziz, M. (2019). Changes needed in Saudi universities' mathematics curricula to satisfy the requirements of vision 2030. *EURASIA Journal of Mathematics, Science and Technology Education, 15*(12). https://doi.org/10.29333/ejmste/109328

Anderssen, B., Broadbridge, P., Fukumoto, Y., Kamiyama, N., Mizoguchi, Y., Polthier, K., & Saeki, O. (2016). *The role and importance of mathematics in innovation: Proceedings of the forum 'math-for-industry' 2015.* Springer.

Blaken, L. (1986). Babylonian astronomers have inspired great minds from Chaucer to Einstein. *Humanities, 7*(5), 11–12.

Clagett, M. (1999). *Ancient Egyptian science: A source book, volume three – Ancient Egyptian Mathematics.* American Philosophical Society.

Dooley, T., & Corcoran, D. (2007). Mathematics: A subject of rights and wrongs? In P. Downes & A. L. Gillingan (Eds.), *Beyond educational disadvantage* (pp. 216–228). Institute of Public Administration.

Dossey, J. A. (1992). The nature of mathematics: Its role and its influence. In D. A. Grouws (Ed.), *Handbook of Research on Mathematics* (pp. 39–48). Macmillan.

Fisher, C. (1990). The research agenda project as prologue. *Journal for Research in Mathematics Education, 21,* 81–89.

Fomunyam, K. G. (2020). *Theorizing STEM education in the 21st century.* IntechOpen.

Formaggia, L. (2017). 'Mathematics and industry 4.0', International CAE Conference,. Retrieved from: https://www.researchgate.net/publication/321155366_Mathematics_and_Industry_40

Fried, M., & Amit, M. (2005). A spiral task as a model for in-service teacher education. *Journal of Mathematics Teacher Education, 8,* 419–436.

Hovarth, J., & Cameron, R. (2017). *3D printed science projects volume 2: Physics, math, engineering and geology models.* Apress Berkeley.

Ikeda, S., & Kotani, M. (2015). A new direction in mathematics for material science. In *Springer briefs in the mathematics of materials book 1.* Springer Japan.

Khalil, N. M., & Osman, K. (2017). *STEM – 21cs module: Fostering 21st century skills through integrated STEM', K-12* (pp. 225–233). STEM Education.

Knorr, W. R., Fraser, C. G., Gray, J. J., Berggen, J. L., & Folkerts, M. (2020, November 9). Mathematics. Retrieved from Encyclopedia Britannica: https://www.britannica.com/science/mathematics/

Leoste, J., & Heidmets, M. (2019). The impact of educational robots as learning tools on mathematics learning outcomes in basic education. In *Digital turn in schools—Research, policy, practice* (pp. 203–217). Springer.

Odell, M. R., & Pedersen, J. L. (2020). Project and problem-based teaching and learning. In B. Akpan & T. Kennedy (Eds.), *Science education in theory and practice* (pp. 343–357). Springer Nature.

Pleasants, J., Clough, M. P., & Olson, J. K. (2019). The urgent need to address the nature of technology: Implications for science education. In B. Akpan (Ed.), *Science education: Visions of the future* (pp. 31–45). Next Generation Education.

Ramful, A., & Patahuddin, S. M. (2021). The fourth industrial revolution: Implications for school mathematics. In J. Naidoo (Ed.), *Teaching and learning in the 21st century: Embracing the fourth industrial revolution* (pp. 13–29). Brill Sense.

Ronis, D. L. (2008). *Problem-based learning for math and science: Integrating inquiry and the internet*. Corwin Press.

Siekmann, G., & P.K. (2016). *Defining 'STEM' skills: Review and synthesis of the literature*. NCVER.

Trilling, B., & Fadel, C. (2009). *21st century skills learning for life in our times*. Wiley.

Violatti, C. (2013). *Ancient greek science*. Retrieved from: World history Encyclopedia: https://www.worldhistory.org/Greek_Science/

Xie, Y., Fang, M., & Shaunman, K. (2015). STEM education. *The Annual Review of Sociology, 41*(19), 1–19.

Yates, C. (2017). *Five ways ancient India changed the world with maths*. Retrieved from: *The Conversation:* https://theconversation.com/five-ways-ancient-india-changed-the-world-with-maths-84332#:~:text=As%20well%20as%20giving%20us,was%20first%20seen%20in%20 India.

Ajeevsing Bholoa is a Senior Lecturer and currently the Head of the Mathematics Education Department at the Mauritius Institute of Education. He has been closely involved in the development of school mathematics curriculum at the pre-primary, primary and secondary levels and the development of courses for the training of mathematics educators. His research interests are related to integration of technology, pedagogical content knowledge and the teaching and learning of mathematics.

Ajay Ramful is an Associate Professor and lecturer in the Mathematics Education Department. He is also co-ordinating the activities of the curriculum development at the Mauritius Institute of Education. He has published widely, including contributions to *Mathematics Education Research Journal*.

Chapter 8
Language in Science and Technology Education

Metin Sardag, Gokhan Kaya, and Gultekin Cakmakci

Abstract Students should be granted opportunities to practice the language of science. This chapter presents a holistic understanding of language in science and technology education by discussing different frameworks in the field. The chapter emphasizes the role of language in learning and knowledge construction processes. In particular, it focuses on the nature of talk in science classrooms and how different interactions between teachers and students contribute to students' learning. The chapter also highlights classroom practices for how language is used in science and technology activities. Several studies suggest that dialogic conversations and discussions should be encouraged in the classroom. This suggestion leads us to consider teachers' classroom interactional competence and the role of language in assessment.

Keywords Conversation analysis · Discourse analysis · Classroom interactional competence · Sociocultural theory · STEM education

Language Perspective of Vygotsky/Sociocultural Theory

Literature sheds light on the role of language in science education using four approaches: sociocultural theory, conceptual change theory, situated learning, and sociolinguistics (Carlsen, 2007). In this chapter, we mainly present and discuss

M. Sardag (✉)
Van Yuzuncu Yil University, Van, Turkey
e-mail: metinsardag@yyu.edu.tr

G. Kaya
Kastamonu University, Kastamonu, Turkey
e-mail: gkaya@kastamonu.edu.tr

G. Cakmakci
Hacettepe University, Ankara, Turkey
e-mail: cakmakci@hacettepe.edu.tr

© The Author(s), under exclusive license to Springer Nature
Switzerland AG 2023
B. Akpan et al. (eds.), *Contemporary Issues in Science and Technology Education*, Contemporary Trends and Issues in Science Education 56,
https://doi.org/10.1007/978-3-031-24259-5_8

language in science and technology education in terms of sociocultural theory, because sociocultural theory clearly defines that the interaction process between the learner and others plays an essential role in learning. As Lijnse (1995) puts it: *'To be able to build on students' knowledge, and to use their constructions productively, we should first know what they really mean when they say what they say'* (p.193). The theory is essential in terms of classroom interactions, as it is also used in the meaning-making process (Mortimer & Scott, 2003) between student-student and student-teacher. Lev Semyonovich Vygotsky was a psychologist who developed sociocultural theory by investigating the conceptualization of learning and human development (Vygotsky, 1978). His theory can be identified in three dimensions: (1) *social origins of human mental development; (2) the role of social interaction in this process; and (3) the primacy of cultural mediation'* (Evnitskaya, 2012, p.16). In line with these dimensions, the social environment, especially the *'external social environment'* where the person's life is developed, is an essential factor that should be examined and analyzed (Jaramillo, 1996), since the learning of an individual necessitates a particular social environment and a social process for bringing up the child in this environment (Vygotsky, 1978). For instance, social environments, where learning occurs through interaction, can vary from doctor-patient interaction in a hospital to teacher-student interaction in a classroom or judge-lawyer interaction in a courtyard regarding interactants and their roles. Vygotsky states that two people with the same level of Intelligence Quotient (IQ) would have different degrees of learning because of their different developments caused by the differences in their social environments. Therefore, learning is defined *'as a social activity like others such as reading a book or listening to music; activities which have an inseparable social dimension whether performed alone or with others'* (Walsh, 2006, p.33). Furthermore, Vygotsky emphasizes that learning includes the mental process that is inextricably linked to our social identity and cognitive schema. Thus, learning is provided neither by cognitive processes nor social interaction alone, but through an interwoven structure for human learning.

The relation between social environment and learning reveals some interactional components. Language, for instance, emerges as a means of communication between the child and the surroundings as the most valuable psychological tool (Vygotsky, 1978, p.89). As given among the fundamental features of sociocultural theory, a teacher should constitute a social plane with which to work with students in a classroom setting, as with friends talking in a restaurant. In this social event or plane, every participant has a chance to reflect on their own understanding, and those reflections create meaning-making processes. Language use, gestures and images are used in social exchanges to express individual thinking as mediator tools during the whole process. There is a transition from social to individual planes throughout the process, as this is strongly emphasized from a sociocultural perspective. The transition is also an indicator of cognition of learning. The tools that provide changes in communication to facilitate internalization also provide means for individual thinking. Consequently, this explains the relationship between thinking and communication, called *'thinking and speech'* by Vygotsky (1934). Mortimer and Scott (2003) refer to the relationship between thinking and communication as important in interaction, as follows:

> *'The intimate relationship between talking and thinking becomes very apparent when we start "talking to ourselves" or "thinking aloud" about difficult, or stressful, problems'* (p.10).

As stated above, learning is seen as a process of internalization in sociocultural theory. Nevertheless, internalization is unimaginable elsewhere than the social plane, because learning is a movement that goes from analysed social to individual. Consistent with the sociocultural perspective, if someone aims to investigate how people tend to think about the world around them, they should initially explore how people talk and communicate around the world, or in a special context such as a classroom. To this end, this chapter explains the talk and communication in science and technology classrooms to find out how learning and teaching emerge in these classrooms.

Language and Interaction in Science and Technology Classrooms

Regarding the tremendous impact on understanding human learning and development, it can be expressed the sociocultural perspective has provided significant implications for educational psychology and other educational research agendas. Teachers or others in a knowledgeable position have a vital role in enabling the social plane, which is referred to by Vygotsky as an existing ground of interactions for knowledge transfer from a social environment to an individual. These teaching and learning perspectives provided the basis for many kinds of research in educational studies, including a great variety of research disciplines such as second language acquisition (e.g., Walsh, 2006) and science education (e.g., Kaya et al., 2016), as well as education research in general.

The sociocultural theory also focuses on the distance or 'cognitive gap' that exists between what people can do individually and what they can do in co-operation with a more skilled other. This cognitive gap is defined as a Zone of Proximal Development (ZPD), which is the distance between the actual developmental level determined by individual problem-solving and the level of potential development determined through problem-solving under adult guidance or in collaboration with more capable peers (Vygotsky, 1978). Lantolf (2000) defines ZPD as a *'collaborative construction of opportunities for individuals to develop their mental abilities'* (p.18). Thus, 'collaboration' is the critical element for the ZPD. The collaborative construction of possibilities for learning is examined through how teachers and learners collectively construct meaning in science classroom interactions. These classroom interaction processes provide the co-construction of knowledge as a joint effort.

Bakhtin (1981), a follower of Vygotsky, greatly influences the educational implications of the sociocultural theory. Bakhtin emphasizes *'the role of social life and of
"the other" informing individual consciousness'* (Mortimer & Scott, 2003, p.12).
Therefore, Bakhtin mainly focuses on the dialogic process, which is an essential
feature for the emergence of learning between expert and novice learners. The process of learning is not directly to transfer knowledge from teacher to students or
adults to children. For Bakhtin, existence, language and thinking were essentially
through a dialogue. Accordingly, it can be asserted that the key points of Bakhtin's
perspective concern the dialogic nature of understanding. In brief, the meaning-
making process, or learning, emerges during the dialogic process between teacher-
student, parents-children, or friend-friend. To this end, more effective classroom
talks or classroom interactions in which there are discussions or decision-making
processes for carrying out investigations may affect the quality of learning or the
teaching environment (Kaya & Cakmakci, 2021; Sardag & Cakmakci, 2021).

Different meanings are generated in a classroom context through teacher and
student interactions. According to the Bakhtinian dialogic process and Vygotskyan
sociocultural theory, teachers and students bring together and work on ideas in
interaction. In line with this understanding, Mortimer and Scott (2003) state that
teaching science involves introducing the learner to the social language of school
science. For this reason, the teacher is at the center of the teaching and learning
process that occurs through the dialogic process. Therefore, their role can be identified as the interpreter or the mediator in the classroom.

According to Mortimer and Scott (2003), science teaching must involve three
fundamental parts.

First, the teacher must make scientific ideas available on the social plane of the
classroom. Second, the teacher needs to assist students in making sense of and inter-
nalizing those thoughts. Finally, the teacher needs to support students in applying
the scientific ideas, while gradually handing over to the students' responsibility for
their use (p.17).

Teaching, learning and doing science are all social processes that are taught and
learned, and they raise individuals who belong to small, such as a classroom, or
larger, social communities (Lemke, 1990). The most important factor is interaction
in this social process. In such a social community as a classroom, teacher-student
and student-student interaction are crucial factors. Lemke was the first researcher to
emphasize the importance of classroom interaction in science teaching. He used the
term 'talking science', which does not only refer to talking about scientific subjects.
Talking science uses language when someone works away from science (Lemke,
1990), which refers to the use of the language of science in and through observation,
classification, analysis, debate, forming a hypothesis, designing an experiment, and
so on. Lemke *(ibid.)* also highlights the importance of the teacher's role in eliciting
'talking science'. He states that teachers' talk shapes how talking science occurs in
classroom settings. Some research that investigates talking science through classroom interaction and teacher-student interaction points to differences in terminology. Lemke (1990) describes classroom interaction as a triadic pattern through

'question-answer-evaluation', while Cazden (2001) refers to *'initiation-response-evaluation'*. Mehan (1979) and Sinclair and Coulthard (1975), on the other hand, focus on the *'initiation-response-follow up'* pattern. Nowadays, the fundamental structure has been mostly kept unchanged, but the general pattern was expanded by Scott et al. (2006). They defined classroom interaction as a chain structure that unfolds through *'initiation-response-feedback-response-feedback'*.

The Role of Language in Assessment Practices in Science and Technology Education

Under this heading, we underline the role of language in assessment practices, presenting assessment types and the understanding and ways of their implementation. In particular, we clarify the role of language in formative assessment (FA). The primary purpose of FA is *'assessment for learning'*, and the second purpose is *'assessment of learning'* (Bennett, 2011). Although the purposes of FA are clear, it is challenging to explain what FA is. Parallel to this, many definitions of FA have emerged and ambiguities and misunderstandings revealed about it. To resolve this, the third International Conference on Assessment for Learning was carried out in 2009. Consequently, the definition that reached consensus in the conference is: *'Assessment for Learning is part of everyday practice by students, teachers and peers, which seeks, reflects upon and responds to information from dialogue, demonstration and observation in ways that enhance ongoing learning'* (Klenowski, 2009, p.264).

Shavelson, Yin, et al. (2008) treat formative assessment as a process ranging from informal to formal formative assessment. The state of the formative assessment might vary depending on the formality of the technique used and the nature of the feedback given to the student by the teacher. In this context, Shavelson, Yin, et al. (2008) focused on three crucial formative assessment techniques. These are *'on-the-fly formative assessment'*, *'planned-for-interaction formative assessment'* and *'embedded-in-the-curriculum formative assessment'* techniques. It is seen that *'on-the-fly formative assessment'* is also considered as an assessment conversation (Duschl & Gitomer, 1997) or informal formative assessment (Ruiz-Primo & Furtak, 2006). On-the-fly formative assessment reveals when an unexpectedly teachable moment occurs (Shavelson, Young, et al., 2008). To close the gap in student comprehension and give appropriate feedback to students, this provides opportunities via learning about students' level of comprehension (Furtak & Ruiz-Primo, 2008) in one-on-one small groups, or whole-group discussion interactions. If we interpret this in the context of classroom interaction in which language is used as a mediator, it is seen that teacher-student interaction and observed student-student interaction provide opportunities for teachers to gain and interpret actual understanding of students, and shape their lesson flow in terms of reached decisions for achieving learning goals. In addition, a teacher's use of languages, such as creating interactional space and instructional idiolect, as we explain below, creates the basis for the interaction that the teacher will establish.

The planned-for-interaction formative assessment is preconceived, as opposed to an on-the-fly formative assessment. Teachers plan to learn about the difference between what the student knows and what they need to know (Shavelson, Yin, et al., 2008) and decide how to bring out learning evidence during teaching. While developing a lesson plan, a teacher can prepare several central questions that touch on the critical points of the learning objectives of that day's course to obtain students' ideas. At the right moment in the course, the teacher allows students to present and discuss their ideas by asking these questions and learning what they know (Shavelson, Yin, et al., 2008). If a teacher tries to use and shape the lesson in light of obtained data, we can say that the teacher transitions from the planned-for-interaction formative assessment technique to the on-the-fly formative assessment.

The embedded-in-the-curriculum formative assessment is an assessment placed in the curriculum in a more formal structure to create moments that can be taught deliberately by teachers or those who develop the curriculum. These assessments are placed at transition points or specific locations in a unit. The assessments provide teachers with information about what students currently know and what they still need to learn so that teachers can give feedback to students at an appropriate time. The embedded-in-the-curriculum formative assessment is more advanced than the other two formative assessment techniques (Shavelson, Yin, et al., 2008).

Teacher's Classroom Interactional Competence in Science and Technology Classrooms

Classrooms are complex institutional environments in which students' needs, the diversity of these, and teachers' abilities vary. Depending on these, learning and understanding are affected by many factors that can create opportunities, or may cause negative outcomes. In this part, since we underline interactional situations, we try to highlight the factors that support learning and enhance the quality of science and technology classroom interaction in terms of teachers. In this understanding, we reflect on the understanding of 'classroom interactional competence' (CIC), which is conceptualized by Steve Walsh (2006, 2011). Walsh (2006) draws on language classrooms to identify these competencies, but these competencies are not context-bound and can be used for different contexts. These competencies illustrate that effective, learning-oriented interaction is more than the quality of teacher talk. With that in mind, we have used science classroom examples such as inquiry-based learning and argumentation-based learning, or STEM education activities, to explain these competencies in this chapter.

Walsh (2011) defines CIC as *'teachers' and learners' ability to use interaction as a tool for mediating and assisting learning'* (p.158). In this regard, it might be said that CIC covers the features of classroom interaction and the competency of teachers and students that cause learning/teaching processes to be more or less effective (Sert, 2015). These features are that: (a) CIC facilitates interactional spaces; (b) CIC shapes learner contributions; (c) CIC makes effective use of eliciting; (d) instructional idiolect; and (e) interactional awareness (Walsh, 2006).

We have reflected on these features for science and technology education classrooms, with extracts to make them more understandable.

CIC Facilitates Interactional Spaces

Interactional space might be increased by several interactional tools, which are: increased wait-time, promoting extended student's turns, allowing planning time, and avoiding the filling of silences (Walsh, 2006). A pioneering study on increased wait time, carried out by Rowe (1974), highlights the influence of wait time on students and teachers. The study points out mainly that when teachers increased wait time: (a) the length of responses produced by students increased; (b) students produced a greater number of appropriate responses, and (c) lower-attaining students' contributions and the incidence of speculative thinking increased. Facilitating the interactional spaces may offer the opportunity to increase learner contribution (Kaya, 2017), since students have the chance to organize their ideas and planning processes.

CIC Shapes Learner Contributions

Some interactional strategies, such as seeking clarification, scaffolding, modeling, or repairing learning input, enable teachers to shape learner contributions (Walsh, 2006). Teachers frequently perform these strategies in an informal formative assessment process (Sardag, 2019).

CIC Makes Effective Use of Eliciting

The ability to use some eliciting strategies and recognize their function is an essential feature of CIC (Walsh, 2006). Teachers frequently ask questions in order to manage interactional processes and reach lesson goals (Sardag, 2019). However, teachers face difficulties in asking appropriate questions to reveal and support students' understanding and learning (McNeill & Knight, 2013). This affects the quality of classroom interaction to elicit students' understanding of formative assessment, as well as the co-construction process. We present Extract 8.1 to show how effectively the use of eliciting deals with the aforementioned issue.

As can be understood from the interaction in Extract 8.1, the teacher asks some referential questions and tries to elicit students' ideas and activities (lines 1, 6, 11–12). These interactional activities related to the teacher's CIC provide students with the opportunities to express their ideas and produce extended learner turns. Additionally, the teacher gains an understanding of students' current levels or positions in the activity process and shapes the interaction.

Extract 8.1 Students Work Together to Produce a New Lighting System to Save Electrical Energy in a STEM Activity

1	Tchr:	What are you doing?
2	Stds:	Teacher, we're going to put the aluminium foil here and
3		Make the light reflect from here.
4	Tchr:	Hmm, I mean…You want the light on the roads.
5	Stds:	Yes.
6	Tchr:	How is it going to be? Do you think it will be useful?
7	Stds:	So, teacher. Since the light will reflect from the bottom,
8		There must be something reflective like aluminium foil,
9		That is, a mirror. And when it is like a
10		Mirror, the light will be reflected there and only on the
11		Road. The light won't go upwards.
12	Tchr:	Well, let me ask you something. How accurate would it be
13		For us to use light on the roads? How safe would it be?

Extract 8.2

Teacher 1: *[…] element of my own individual conversational style or lack of it that are carried into the classroom, engrained habits that I would have to take a crowbar to prise out of myself. Or maybe wouldn't want to remove, but maybe it's a good idea to be aware of them, that they can take over or that they can sometimes not be the most constructive approach.*

In this example, the teacher clearly explains that her habits affect the class, and she states that she is aware of these habits and must remove them. There may be other behaviours that are overused by teachers and which affect the interaction structure. For example, some are habits such as expressions of approval that the teacher uses excessively, pauses, or speaking before the students' answers are finished (Kaya et al., 2016).

Instructional Idiolect

Instructional idiolect reflects teachers' speech habits that may support the construct of learning opportunities, or cause obstacles for them (Walsh, 2006). The idiolect could be related to their lives outside the classroom life, their regional accent, or some stereotypical teaching styles, and these all affect their speech habits. To exemplify the instructional idiolect, we have drawn on Extract 8.2, taken from Walsh's (2006, p.139) book, which is related to the reflective process of teacher talk. In that case, the teacher comments on the features of her everyday talk that had been 'carried' into the classroom.

Interactional Awareness

Interactional awareness is at the centre of CIC, which is evidenced through many data (Walsh, 2006). The level of interactional awareness might change both from teacher to teacher, and from the moment in which teaching occurs to another moment. This situation can be explained through teachers' sensitivity to their role at a particular moment or stage of the lesson (Walsh, 2006).

Summary

This chapter discusses language and learning perspectives, which have an essential understanding of the sociocultural theory of Vygotsky and science and technology classes in language and learning contexts. In addition, assessment processes where learning and interaction are at the forefront, informal formative assessment, and language-in-use, as well as some of the abilities that construct teachers' interactional competence emphasized in order to transition from theory to practice in light of the examples of science and technology classroom interaction.

Recommended Resources

Carlsen, W. S. (2007). Language and science learning. In Abell, S. K. & Lederman, N. G. (Eds.), *Handbook of research on science education* (pp. 57–74). Lawrence Erlbaum.

Lemke, J. L. (1990). *Talking science: Language, learning and values*. Ablex Publishing.

Mortimer, E., & Scott, P. (2003). *Meaning making in secondary science classrooms*. McGraw-Hill Education.

Walsh, S. (2006). *Investigating classroom discourse*. Routledge.

References

Bakhtin, M. M. (1981). *The dialogic imagination: Four essays*. University of Texas Press.

Bennett, R. E. (2011). Formative assessment: A critical review. *Assessment in Education: Principles, Policy & Practice, 18*(1), 5–25. https://doi.org/10.1080/0969594X.2010.513678

Carlsen, W. S. (2007). Language and science learning. In S. K. Abell & N. G. Lederman (Eds.), *Handbook of research on science education* (pp. 57–74). Lawrence Erlbaum.

Cazden, C. B. (2001). *Classroom discourse: The language of teaching and learning*. Heinemann.

Duschl, R., & Gitomer, D. H. (1997). Strategies and challenges to changing the focus of assessment and instruction in science classrooms. *Educational Assessment, 4*(1), 37–73. https://doi.org/10.1207/s15326977ea0401_2

Evnitskaya, N. (2012). *Talking science in a second language: The interactional co-construction of dialogic explanations in the CLIL science classroom.* (unpublished doctoral dissertation). Universitat Autònoma de Barcelona.

Furtak, E. M., & Ruiz-Primo, M. A. (2008). Making students' thinking explicit in writing and discussion: An analysis of formative assessment prompts. *Science Education, 92*(5), 799–824. https://doi.org/10.1002/sce.20270

Jaramillo, J. (1996). Vygotsky's sociocultural theory and contributions to the development of constructivist curricula. *Education, 117*(1), 133–140.

Kaya, G. (2017). *Teacher talk and learner contributions in inquiry-based science education: A conversation analytic examination.* (unpublished doctoral thesis). Hacettepe University.

Kaya, G., & Cakmakci, G. (2021). Conversation analytic examination of inquiry-based science classrooms. *Kastamonu Education Journal, 29*(3), 736–755. https://doi.org/10.24106/kefdergi.940307

Kaya, G., Sardag, M., Cakmakci, G., Doğan, N., İrez, S., & Yalaki, Y. (2016). Discourse patterns and communicative approaches for teaching nature of science. *Education and Science, 41*(185), 83–99. https://doi.org/10.15390/EB.2016.4852

Klenowski, V. (2009). Assessment for learning revisited: An Asia-Pacific perspective. *Assessment in Education: Principles, Policy & Practice, 16*(3), 263–268. https://doi.org/10.1080/09695940903319646

Lantolf, J. P. (2000). *Sociocultural theory and second language learning.* Oxford University Press.

Lemke, J. L. (1990). *Talking science: Language, learning and values.* Ablex Publishing.

Lijnse, P. L. (1995). "Developmental research" as a way to an empirically based "didactic structure" of science. *Science Education, 79*(2), 189–199.

McNeill, K. L., & Knight, A. M. (2013). Teachers' pedagogical content knowledge of scientific argumentation: The impact of professional development on K-12 teachers. *Science Education, 97*(6), 936–972. https://doi.org/10.1002/sce.21081

Mehan, H. (1979). *Learning lessons: Social organization in the classroom.* Harvard University Press.

Mortimer, E., & Scott, P. (2003). *Meaning making in secondary science classrooms.* McGraw-Hill Education.

Rowe, M. B. (1974). Wait-time and rewards as instructional variables, their influence on language, logic and fate control: Part one-wait-time. *Journal of Research in Science Teaching, 11*(2), 263–279. https://doi.org/10.1002/tea.3660110202

Ruiz-Primo, M. A., & Furtak, E. M. (2006). Informal formative assessment and scientific inquiry: Exploring teachers' practices and student learning. *Educational Assessment, 11*(3–4), 237–263. https://doi.org/10.1080/10627197.2006.9652991

Sardag, M. (2019). *Formative assessment in argumentation-based science education: A conversation analytic research.* (unpublished doctoral thesis). Hacettepe University.

Sardag, M., & Cakmakci, G. (2021). Interactional resources and teachers' questions in argumentation-based science education. *YYU Journal of Education Faculty, 18*(2), 494–523. https://doi.org/10.33711/yyuefd.1029064

Scott, P. H., Mortimer, E. F., & Aguiar, O. G. (2006). The tension between authoritative and dialogic discourse: A fundamental characteristic of meaning making interactions in high school science lessons. *Science Education, 90*(4), 605–631. https://doi.org/10.1002/sce.20131

Sert, O. (2015). *Social interaction and L2 classroom discourse.* Edinburgh University Press.

Shavelson, R. J., Yin, Y., Furtak, E. M., Ruiz-Primo, M. A., Ayala, C. C., Young, D. B., Tomita, M. K., Brandon, P. R., & Pottenger, F. (2008). On the role and impact of formative assessment on science inquiry teaching and learning. In J. Coffey, R. Douglas, & C. Stearns (Eds.), *Assessing science learning: Perspectives from research and practices* (pp. 21–36). NSTA Press.

Shavelson, R. J., Young, D. B., Ayala, C. C., Brandon, P. R., Furtak, E. M., Ruiz-Primo, M. A., Tomita, M. K., & Yin, Y. (2008). On the impact of curriculum-embedded formative assessment on learning: A collaboration between curriculum and assessment developers. *Applied Measurement in Education, 21*(4), 295–314. https://doi.org/10.1080/08957340802347647

Sinclair, J., & Coulthard, M. (1975). *Towards an analysis of discourse.* Oxford University Press.

Vygotsky, L. S. [1934, 1935, 1962] (1978). *Mind in society: The development of higher mental processes*. In M. Cole, V. John-Steiner, S. Scribner & E. Souberman (Eds.), Reprint. Harvard University Press.

Walsh, S. (2006). *Investigating classroom discourse*. Routledge.

Walsh, S. (2011). *Exploring classroom discourse: Language in action*. Routledge.

Metin Sardag holds a Master of Arts and a PhD degree in Science Education. He is an Associate Professor of Science Education at Van Yuzuncu Yil University. He has been teaching courses on argumentation education at the tertiary level. His research focuses on classroom interaction, argumentation, and informal formative assessment in science classrooms. He has been involved in some national and international projects.

Gokhan Kaya holds a Master of Arts and PhD degrees in Science Education from Hacettepe University. He is an Associate Professor of Science Education at Kastamonu University. His research interests focus on classroom interaction among students and teacher-student and classroom discourse in the science classroom. He has worked as a researcher in many European projects related to inquiry based learning and teacher education.

Gultekin Cakmakci is a Professor of Science Education at Hacettepe University and has been teaching courses on STEM education and public engagement with STEM. His research interests focus on developing scientific literacy among students and the general public and on the design, implementation and evaluation of STEM teaching. He is currently a board member of the *European Science Education Research Association (ESERA)*, and *Journal of Research in STEM Education*.

Chapter 9
The Real and Virtual Science Laboratories

Shakeel Mohammad Cassam Atchia ⓘ and Anwar Rumjaun ⓘ

Abstract Laboratory-based practical work forms an integral part of science education, as it does not only confer opportunities to develop scientific knowledge, but it also supports the development of science process skills, attitudes and values, which are key to face current and future challenges. However, many education systems face challenges to embed real practical work in the teaching and learning of science. This is due to several reasons, including lack of infrastructure and resources. Could virtual laboratory experimentation be an alternative way to address these challenges and make practical work a regular activity in science education, without compromising the quality of students' engagement in practical work?

It is in this perspective that this chapter (1) explores some of the initiatives taken towards this shift; (2) provides an overview of the opportunities, limitations and challenges associated with virtual laboratory-based practical work; and (3) situates the debate on the extent that virtual laboratory-based practical work can substitute for the real laboratory-based practical work in providing effective learning environments suitable to students' needs in this digital age.

Keywords Real laboratory-based practical work · Virtual laboratory · Science education · Scientific skills and competencies · Debate and future of practical work

Introduction

Drawing on the nature of science and the socio-constructivist pedagogical approach, it is undeniable that laboratory-based practical work forms an integral part of science education, as it does not only confer opportunities to develop scientific knowledge, but also supports the development of science process skills, attitudes and

S. M. C. Atchia · A. Rumjaun (✉)
Mauritius Institute of Education, Moka, Mauritius
e-mail: Shakeel.Atchia@mie.ac.mu; a.rumjaun@mie.ac.mu

© The Author(s), under exclusive license to Springer Nature   113
Switzerland AG 2023
B. Akpan et al. (eds.), *Contemporary Issues in Science and Technology Education*, Contemporary Trends and Issues in Science Education 56,
https://doi.org/10.1007/978-3-031-24259-5_9

values, which are key to face current and future challenges. However, in today's context, coupled with the current COVID-19 pandemic and its associated school closures, many education systems around the world are reviewing their priorities and shifting towards distance education to maintain the continuity of teaching and learning in all curriculum areas, including science.

Though the shift towards remote and online teaching and learning is incontestably a laudable initiative, science by the nature of its discipline requires students' engagement in inquiry learning in laboratory and field work, thus presenting a dire need to reflect on the approach to be adopted, and on the role that the virtual laboratory (VLab) will play in providing effective learning environments suitable to students' needs and science education in this digital age.

This chapter will therefore focus on some espoused initiatives related to this shift and explore the opportunities, limitations and challenges that real (hands-on) and virtual laboratory-based practical work can offer, especially at a time when COVID-19 has curtailed access to real school laboratories in different parts of the world due to schools' closures. This chapter will also situate the debate on the effect of VLab experiments on students' learning by drawing on the scholarship around the importance, types, modes and effects of VLab on the meaningful teaching and learning of science.

Real Lab-Based Practical Work in Science

This section provides an overview of the real lab-based (RLab) practical, before unpacking the debate around the use of virtual lab-based practical work in science education. It focuses on the role and purpose, types, underpinning learning theories and assessment of lab-based practical work.

Role and Purpose of Lab-Based Practical Work in Science

Over the past decades, the dominant epistemological view of science education has gradually shifted, from an inductive to a hypothetico-deductive paradigm, where the focus is on the engagement of students in the construction of knowledge and development of scientific skills, attitudes and values. Shaping of this avowed shift remains incomplete without students' engagement in practical work. In fact, RLab practical plays important roles in the teaching and learning of science at all levels, from pre-primary to tertiary, independent of the educational settings. It serves as a strategy to develop and consolidate understanding of the scientific concepts, in addition to its application in different experimental situations. In fact, the teaching and learning of science remains incomplete without students' engagement in practical work to construct increasingly sophisticated and powerful representations of the world.

Practical work has a much more important and valuable role in science education when it is integrated in inquiry or problem-solving activities to help students' construct scientific knowledge and understanding. In such cases, students are given the leeway to identify issues, set hypotheses, identify research questions, design and implement experiments, collect, present and analyse data, to eventually infer and conclude. Practical work in this setting not only serves the purpose of constructing meaningful knowledge, but also allow students to develop key and higher-order scientific inquiry skills, attitudes and values.

Types of Lab-Based Practical Work

The typological classification of laboratory-based practical work is subjective and cannot be formalized due to its classification regarding disparate aspects of practical work documented in the literature. For instance, Woolnough and Allsop (1985) and the National Research Council (1996) categorized practical work based on its purpose, namely, *exercises, investigations, experiences* and *demonstration*. Several other researchers have categorized practical work based on the level of openness and the demand for inquiry skills, producing a four-way classification of practical work, depending on whether each stage is open, that is, left to the students to decide, or closed (Tamir, 1991). At level zero, all the problems, procedures and data are given to enable students to draw conclusions and, hence, there is no experience of scientific inquiry. In level one, both problems and procedures are given, and students collect and analyze data to draw relevant conclusions. In level two, only the problem is given, and the students must design the procedure, collect the data, and draw conclusions. In level three, the students must do everything, that is, from problem formulation up to drawing of conclusions.

Yet another, more recent and context-relevant, classification was given by Shimba et al. (2017), where lab-based practical work was classified into hands-on and virtual. In hands-on, students are physically present in the lab to manipulate equipment, apparatus and reagents, whereas in virtual, all manipulation is made through computer-generated and simulated platforms.

Learning Theories Supporting Lab-Based Practical Work

This section focuses on some key learning theories underpinning practical work, namely, experiential, behaviorism, cognitivism, humanism, constructivism, and multiple intelligences learning theories.

According to Kolb's theory (1984), experiential learning is a holistic process that works in four stages, namely the concrete learning, reflective observation, abstract conceptualization and active experimentation, which are also key components of lab-based practical work. In fact, it has been rightly stated that practical work provides strategic, active engagement of students in opportunities to learn through doing, and reflection on those activities, which empowers them to apply their theoretical knowledge to practical endeavors in a multitude of settings.

Behaviorism (Skinner, 1968), as a learning theory, studies observable and measurable behavioral changes based on stimulus-response associations, centered around students' abilities to counteract any incidence or situations cropping up during implementation of practical tasks. Writing hypotheses, calculating responses, focusing a microscope, balancing a scale, weighing pebbles, and keeping track of observations are all examples of active reacting and thus responses are based on the stimulus-response mechanism described by the behaviorism learning theory.

Cognitivism, developed by Wolfgang Kohler in the 1900s, relates to the fact that learning happens when the memory system acts as an informative organizer and processor to construct new knowledge based on prior information (Ertmer & Newby, 1993). In fact, lab-based practical work, with a clear-cut path towards understanding how learning should be guided and organized, basically leads towards discovering new knowledge from previous experiences, which may be corrected, consolidated or enhanced.

Moreover, the humanistic learning theory developed by Maslow in the 1900s centers around the use of human-specific capabilities such as creativity, individual cognitive growth, decision-making, social skills, feelings, intellect, artistic skills and practical skills in the process of learning. In fact, practical work and the humanistic theories intertwine with the human psychological and cognitive capacities to create learning in a social setting, which is further supported by Vygotsky's theory of 'socio-constructivism', inferring that learning should provoke self-actualization of knowledge. The humanism approach considers the learners' choice, intrinsic motivation, self-evaluation and feelings in a safe environment (Zucca-Scott, 2010).

Constructivism learning theory, which focuses on knowledge construction, is the driver of practical work that uses inquiry as the vehicle. For instance, students construct knowledge and develop skills when engaged in: (1) setting hypotheses and research questions; (2) identifying independent, dependent and controlled variables; (3) designing, planning and implementing relevant investigations; (4) planning required safety measures; (5) generating, recording and analyzing data; (6) identifying sources of errors; (7) modifying procedures to cater for unexpected incidences and errors; and (8) inferring, concluding and recommending future actions. Practical work, using the constructivist approach, not only allows students to construct knowledge based on prior knowledge and past experiences, but also allows students to clear their preconceptions and misconceptions, which often interfere with the ability to learn new material. Practical work is a perfect tool to create an internal conflict, known as cognitive dissonance, as lab work is designed to challenge students' current knowledge so that new knowledge is created. As far as 'learning with a social component' is concerned, practical work provides many opportunities for social interactions, where students discuss their predictions, explanations, procedures and data table before doing the activity, then work in groups to complete the practical, and then present their results and conclusions. The process of formulating an opinion to express and share with a group promotes reflection.

The multiple intelligence theory not only supports hands-on, but also virtual, lab practical work. For instance, a 3D virtual lab provides a simulation environment that allows students to participate in experiments in predefined conditions. The laboratories can be reached through any computer or online network.

Assessment of Lab-Based Practical Work

Laboratory practical is used to assess the quality of knowledge that students have and their abilities to use and apply it in different situations (Sedumedi, 2017). Practical work may also be used to assess one's conceptual understanding, and science process skills (Abrahams et al., 2013). The assessment of practical work allows teachers to gather, synthesize and interpret the performance of individual students to provide a thorough evaluation of their work and deliver appropriate feedback to reinforce learning and ensure student progression. In fact, practical work may be used as a diagnostic, formative or summative assessment tool.

Firstly, laboratory practical work can be used as a diagnostic assessment to find out students' prior knowledge or any misconceptions that they might have pertaining to a specific concept, to ensure a smooth learning procedure. Secondly, practical-based formative assessment is a type of internal assessment that is performed during the learning process to ensure that the required process skills, such as observation, measurement, planning, predicting, experimenting and communication are assessed (Abrahams et al., 2013) in addition to assessing construction of scientific knowledge and development of understandings. Thirdly, practical work may be used as a summative assessment, as it tests whether students are able to apply the knowledge learned and skills developed in a practical situation such as problem-solving, where soft skills, attitudes, communication, critical and observation skills are difficult to assess in written exercises. Therefore, the different types of laboratory practical act as a conspicuous tool, enabling teachers to assess meaningful learning of science.

Current Challenges of RLab-Based Practical Work

RLab-based practical work is one of the fundamental tools that allow students to construct meaningful scientific knowledge and develop skills needed for transformative learning, so that students develop the necessary capabilities to face current and future challenges. However, carrying out practical work in class has its own challenges, such as the associated cost and physical facilities, access to resources and teachers' readiness and reluctance.

As far as cost is concerned, the setting up and maintenance of science laboratories is very expensive. Moreover, specific pedagogical tools based on students' needs, new technology with associated equipment and apparatus, and safety requirements, contribute much to the costs. According to Kasiyo et al. (2017), the main challenges that hinder the practical work affecting science education in many countries is the lack of laboratories in which to conduct experiments, due to the costs associated with setting up a proper lab. The second key challenge is access to resources. Though contextualized, in several countries teachers and students have limited access to resources due to the high cost of resources, political will to invest in quality school labs, and schools situated in remote areas, amongst others. As far as teachers' reluctance and readiness are concerned, some teachers, due to lack of experience, do not feel confident to carry out practical lab classes. Moreover, among the educators who engage students in lab-based practical work, many use the traditional practical class based on the recipe or instruction-based approach. These practicals only improve the manipulative skills of students without developing key inquiry process skills and creating opportunities to think critically, challenge and improve their cognitive development. Traditional laboratory methods failed to introduce the nature of science accurately and, as a consequence of this failure, students tend to accept science as a collection of facts to be memorized rather than a set of scientific principles confirmed by evidence (Alake-Tuenter et al., 2012). Therefore, students do not get a chance to create or influence the work procedure during the lab work. Moreover, many teachers use predetermined experiments in a lab manual, showing that teachers are reluctant to come up with their own practical work. In Mauritius, teachers use lab practical work only for upper-age classes when they have their Cambridge A-level Paper 3 practical examinations. Moreover, these practicals are not used as tools to construct knowledge, but rather as summative practical work.

In addition to the above challenges affecting RLab practical work, the COVID-19-related school closures and the shift from face-to-face to online education remains the prime impact on quality science education.

The Shift from Real to Virtual Lab-Based Practical Work in Science

In the late 1900s, the National Science Teachers' Association (NSTA) in the United States agreed that laboratory experience is *'so integral to the nature of science that it must be included in every science program for every student'* (NSTA, cited in NSTA, 2005, 1). Laboratory-based activities help students to learn concepts, develop inquiry skills and illustrate theory (Johnstone & Alshuaili, 2001). It develops a wide variety of investigative, organizational, creative and problem-solving skills and abilities that are refined in the context of laboratory inquiry. It also increases students' curiosity and positive attitudes toward science (Bretz et al., 2013). In fact, the laboratory is a vital environment in which science is experienced. Despite its importance, lab work is still facing many challenges, as described earlier, such as the high cost of lab equipment and materials, safety issues when dealing with dangerous experiments (Kapting'ei & Rutto, 2014), and teachers' reluctance. This whole debate around the challenges of RLab-based practical activities brings us to a key reflective question: Do we have alternatives to overcome such challenges and to support development of key inquiry and manipulative skills?

According to Sassi (2000) and Kocijancic and Jamsek (2004), there is growing scholarship on the integration of Information and Communication Technologies (ICTs) in science practical work, through virtual laboratories to improve and enhance the development of inquiry skills. Hofstein and Kind (2012) highlighted that the integration of ICT, through simulations, animation, videos and visualizations within real practical work is very promising. In fact, the most employed pedagogical uses of virtual laboratories fall into two apparently distinct categories, which may be classified, respectively, as its use to enhance RLab-based hands-on practical work and as an alternative to practical work that cannot be conducted in RLabs due to lack of equipment and apparatus, safety implications and teachers' reluctance.

What is a Virtual Lab?

A virtual lab (Vlab) can be defined as an online environment that consists of a set of experiment simulations and videos, which allow students to run experiments virtually (Bajpai, 2013) and has the potential to support and enhance face-to-face practical-based learning (Darrah et al., 2014; Sullivan et al., 2017). Students can learn the scientific concepts and gain new skills using virtual lab anytime and anywhere, using only their laptops or even smartphones (Ramesh, 2019). It is a tool that students can use to run their own experiments, using the 'mouse' to control physical actions such as mixing reagents, pipetting solutions, calibrating devices, pushing objects, heating materials and measuring. Animation and simulation concepts are used to allow students to interact with materials and apparatus to see the results of the reaction in an experiment.

Types of Virtual Laboratories

Virtual laboratories are interactive simulations in which students perform experiments, collect data and answer questions to assess their understanding. These types of laboratories have been developed to reproduce experiments that can be carried out in physical laboratories. The central activity in any lab is running experiments and collecting data. Thus, a real virtual lab must include real experiments from which students can collect data for analysis and conclusion (Keller & Keller, 2005).

The list below presents some useful sites for VLabs:

- https://www.ncbionetwork.org/iet/microscope/
- https://learn.genetics.utah.edu/content/labs

- https://praxilabs.com/
- https://amrita.olabs.edu.in
- https://www.biointeractive.org
- https://www.knowitall.org
- https://learn5.open.ac.uk
- https://www.labxchange.org/library
- https://virtuallabs.nmsu.edu
- https://www.labster.com

Advantages and Limitations of VLab Experiments

This section focuses on the advantages and limitations of using Vlab in the teaching and learning of science.

Advantages of Using VLab

VLab provides several advantages such as:

Lower cost and affordable practical work:

Though most virtual labs are not free, VLabs are still less expensive than RLabs. VLab involves the use of one platform that can serve an entire institution without spending on resources and does not require high investment and maintenance costs. Moreover, students have the opportunities to make and learn from their mistakes with minimal negative consequences compared to real labs.

Engagement of students in practicals not feasible in Rlabs:

VLabs allow students' engagement in virtually conducting experiments that real labs will not allow due to lack of equipment, costly materials, dangerous situations, ethical considerations and safety issues. For instance, VLabs provide opportunities to conduct experiments such as dissection of live specimens of plants and animals, which are often associated with ethical considerations. They also allow students to engage in practicals on microbiology without manipulating microorganisms, which is not feasible in Rlabs due to safety issues and lack of the necessary apparatus in secondary school labs.

Flexible access:

All virtual labs are online and thus students have easy access to the platform whenever and wherever they want. They are very useful in the context of pandemic where the sanitary protocols require contactless learning.

Reducing discriminate access:

At higher levels of secondary education, students learn topics such as biochemistry, molecular biology and gene technology, where the relevant practicals therein necessitate equipment and reagents. Schools from the developing countries face huge challenges to procure these resources, as they need to purchase through the relevant authorities from other countries. Therefore, the virtual lab would be a sound and reasonable investment in the long run.

Safety:

In real laboratories, the safety of students and staff is a key consideration. Students are sometimes at risk while manipulating toxic reagents, cancerous substances and corrosive acids, among others. These potential risks are absent while engaged in VLab, though students are informed about all safety procedures and issues.

Space requirement:

Virtual labs do not require space to accommodate students, teachers, technical attendants and all the logistics and resources that should be made available in a real lab space. Thus, a VLab allows every single user to work remotely using their PC/laptop and connectivity.

Limitations of Using VLab

VLabs are mostly focused on experiments whereby all the instructions are laid down and students have to follow and comply with them, including the safety measures, which are also laid down. These experiments provide the students with the opportunity to engage in virtual hands-on activities, but they do not enable students to engage in inquiry and problem-solving as in real-world experimentation. In fact, the National Science Education Standards of the United States highlighted that 'Conducting hands-on science activities does not guarantee inquiry' (National Research Council, 1996, p.23), as the inquiry process is key in enabling students to identify a problem to be solved, reflecting, thinking and asking key questions. Development of such skills is limited in VLab experimentation. With the advent of VLab, the risk of teachers trivializing the inquiry process may be high, as the focus is often on a recipe approach where students are limited to following set instructions. As a result, teachers send their students into the laboratory simply to replicate or verify what has already been explained and illustrated in the classroom, which ultimately leads to an unexciting experience.

Another important limitation of VLab is the complexity of developing VLab experiments and resources that fit curriculum requirements. Although virtual laboratories have become increasingly common as a form of teaching aid in different learning situations (Achuthan et al., 2018), available materials may not fit a particular curriculum and thus require contextualization of resources. However, creating the virtual laboratory is highly complex, incorporating diverse areas such as interaction design, visualization and pedagogy. It involves design and production of texts, images, 3D environments and interactivity, and the production requires programming and animation. The development of a virtual laboratory, as well as implementing it as a laboratory exercise for learning, requires knowledge in the three domains outlined by the TPACK model (Koehler & Mishra, 2009), i.e., technology, pedagogy and content knowledge.

Current Debates Around RLab and VLab

There is ongoing debate between researchers of two schools of thought, where one advocates the use of VLab, while the other argues against its use. In fact, despite such advantages of using VLab as the improvement of accessibility, and the pedagogical advantage of a well-designed virtual laboratory being able to better explain difficult theoretical concepts in the study field concerned, VLabs have often been criticized for their recipe approach to practical work, limiting some key inquiry skills, and the complexity of developing VLab resources. However, it should be noted that, though it is undeniable that most of the currently available VLab practicals are based on the recipe approach, few have been able to integrate the virtual manipulation within inquiry processes. For instance, Feyzioglu (2009) and Prajoko et al. (2016) highlighted that VLab with an inquiry-based learning approach has been positively correlated with the enhancement of the mastery of science process skills (SPS) in some studies. Thus, if properly planned and with appropriate pedagogical supports, VLabs can represent a powerful tool in enhancing science education in this digital era. Moreover, despite the complexity in developing contextualized VLab resources, its importance largely outweighs the limitation.

Based on the debate unpacked in this chapter around the advantages and limitations of RLab and VLab, we believe that students should be offered, through the duration of their programs, a balanced mixture of real and virtual lab experiences. Arguments in favor of using virtual laboratories include accessibility and resource economy. They also cater for demonstration possibilities for things that normally cannot be seen or are difficult to explain (Lewis, 2014). Moreover, Achuthan et al. (2018) showed that learning was improved when students could use virtual labs prior to physical ones. Studies also show that students' learning outcomes are equal, or higher, in non-traditional laboratories, such as virtual laboratories, compared to traditional laboratory environments (Brinson, 2015).

In fact, learning the fundamentals of the topic via a virtual laboratory and then taking advantage of a real, hands-on lab practical might enable students to gain deeper and more complex understandings. Similarly, Zacharia and de Jong (2014) concluded that a virtual laboratory should precede the hands-on laboratory environment. They investigated different combinations of laboratory environments for teaching concepts related to electricity for undergraduate students, and stated that not all hands-on and virtual laboratory environment combinations have the same impact on students' conceptual knowledge. They claimed that a virtual laboratory is more appropriate for acquiring key concepts since it provides instant feedback, which facilitates learning.

VLab is now gaining "terrain" in the STEM field. It has been reported that VLab coupled with a STEM approach, besides improving and promoting the STEM approach in class, helps to develop critical thinking (Trisnaningsih et al., 2021). VLab STEM learning is considered efficient to promote learners' competences of problem solving, decision making and investigative skills (Trúchly et al., 2019). More insights related to STEM learning can be obtained from Chap. 4 entitled

"STEM education as a meta-discipline" and could contribute further to the debate of VLab in STEM learning.

Practical Work in the Future

Development of the virtual laboratory is focused on moving incrementally towards the eventual goal of allowing students to undertake virtual experiments. At present, students can set up the apparatus, manipulate objects, mix reagents, observe under microscopes and perform tasks that they usually do in a real laboratory. The next step is to model molecular simulations, allowing for an experiment to be carried out with the facility to zoom in and visualize processes on a molecular level (Adlong et al., 2003). It is intended to also introduce various symbolic representations. Allowing students to move between macroscopic (laboratory level), microscopic (molecular level) and symbolic representations of science concepts is consistent with research into science pedagogy. When the macroscopic, microscopic and symbolic aspects of science are taught separately, insufficient connections are made between the three levels, and the information remains compartmentalized in long-term memories of students.

The Swedish multi-institutional research program, MultiG, initiated a number of research projects concerned with telecommunication, telecollaboration and telepresence. One of these projects is distributed interactive virtual environments (DIVE), a multi-user virtual reality system developed jointly by the Swedish Institute of Computer Science and the Royal Institute of Technology in Stockholm, Sweden. DIVE is used as the platform for research in collaborative work in virtual spaces. In this model, each participant defines subspaces for their presence and attention. The intersection of those subspaces provides for varying degrees of mutual awareness to support more natural human-human interaction in virtual environments.

Conclusion

No learning system is perfect, and effective learning is achieved using a diverse range of approaches. We therefore tend to argue that school science should provide a blend of learning experiences through virtual labs and traditional hands-on lab experiments. Virtual lab experiment approaches impede the tactile and kinesthetic aspects of a traditional laboratory. Students will not feel, taste or smell the experimental materials.

The challenges of today's science classroom require new solutions. The alternative presented in this chapter uses real experiments, together with interactive data collection, to ensure that the important features of the lab are not skipped. While the use of virtual labs can, to some extent, miss some aspects of a traditional lab, their proper use can compensate for this loss. Students, and most teachers, found it easy to use VLab as an engaging and valuable aid to learning science.

Virtual laboratories cannot replace entirely the physical experiments in traditional laboratories. Using hands-on and virtual laboratories sequentially offers better student performance and hands-on laboratories may have complementary importance (Kapici et al., 2019). For the current context, schools can resort to either type of laboratory experience depending on the prevailing contexts and situations. In situations of pandemic, students can perform the experiments online without any time limitations, receive instant feedback, and familiarize with health and safety regulations.

Summary

In this chapter we have discussed the role and purpose and types of lab-based practical work in science. Furthermore, the learning theories that support lab-based practical work are also presented and discussed. Another key aspect documented in this chapter is the assessment in practical work which is linked to the challenges incurred in implementing RLab-based practical work. The chapter then describes the shift from real to virtual lab-based practical work in science including the types, advantages and disadvantages of virtual laboratories. The debate around RLab and VLab are discussed before providing some insights in the future of practical work.

References

Abrahams, I., Reiss, M. J., & Sharpe, R. M. (2013). The assessment of practical work in school science. *Studies in Science Education, 49*(2), 209–251.

Achuthan, K., Kolil, V. K., & Diwakar, S. (2018). Using virtual laboratories in chemistry classrooms as interactive tools towards modifying alternate conceptions in molecular symmetry. *Education and Information Technologies, 23*(6), 2499–2515.

Adlong, W., Bedgood, D. R., Jr., Bishop, A. G., Dillon, K., Haig, T., Helliwell, S., Pettigrove, M., Prenzler, P. D., Robards, K., & Tuovinen, J. E. (2003). On the path to improving our teaching – Reflection on best practices in teaching chemistry. In *Learning for an unknown future, proceedings of the 2003 Conference of the Higher Education Research and Development Society of Australasia* (pp. 52–60). HERDSA.

Alake-Tuenter, E., Biemans, H. J., Tobi, H., Wals, A. E., Oosterheert, I., & Mulder, M. (2012). Inquiry-based science education competencies of primary school teachers: A literature study and critical review of the American National Science Education Standards. *International Journal of Science Education, 34*(17), 2609–2640.

Bajpai, M. (2013). Developing concepts in physics through virtual lab experiment: An effectiveness study. *Techno LEARN, 3*(1), 43–50. https://ndpublisher.in/admin/issues/tlv3n1f.pdf

Bretz, S., Fay, M., Bruck, L. B., & Towns, M. H. (2013). What faculty interviews reveal about meaningful learning in the undergraduate laboratory. *Journal of Chemical Education, 90*(3), 5–7. https://doi.org/10.1021/ed300384r

Brinson, J. R. (2015). Learning outcome achievement in non-traditional (virtual and remote) versus traditional (hands-on) laboratories: A review of the empirical research. *Computers & Education, 87*, 218–237.

Darrah, M., Humbert, R., Finstein, J., Simon, M., & Hopkins, J. (2014). Are virtual labs as effective as hands-on labs for undergraduate physics? A comparative study at two major universities. *Journal of Science Education and Technology, 23*(6), 803–813. https://doi.org/10.1007/s10956-014-9513-9

Ertmer, P. A., & Newby, T. J. (1993). Behaviorism, cognitivism, constructivism: Comparing critical features from an instructional design perspective. *Performance Improvement Quarterly, 6*(4), 50–72.

Feyzioglu, B. (2009). An investigation of the relationship between science process skills with efficient laboratory use and science achievement in chemistry education. *Journal of Turkish Science Education, 6*(3), 114–132.

Hofstein, A., & Kind, P. M. (2012). Learning in and from science laboratories. In B. Fraser, K. Tobin, & J. M. Campbell (Eds.), *Second international handbook of science education* (pp. 189–207). Springer. https://doi.org/10.1007/978-1-4020-9041-7_15

Johnstone, A., & Alshuaili, A. (2001). Learning in the laboratory: Some thoughts from the literature. *University Chemistry Education, 5*(2), 42–51.

Kapici, H. O., Akcay, H., & de Jong, T. (2019). Using hands-on and virtual laboratories alone or together – Which works better for acquiring knowledge and skills ? *Journal of Science Education and Technology, 28*(3), 231–250.

Kapting'ei, P., & Rutto, D. K. (2014). Challenges facing laboratory practical approach in physics instruction in Kenyan District secondary schools. *International Journal of Advancements in Research & Technology, 3*(8), 13–17.

Kasiyo, C., Denuga, D., & Mukwambo, M. (2017). An investigation and intervention on challenges faced by natural science teachers when conducting practical work in three selected school of Zambezi region in Namibia. *American Scientific Research Journal for Engineering, Technology, and Sciences, 34*(1), 23–33.

Keller, H. E., & Keller, E. E. (2005). Making real virtual labs. *The Science Education Review, 4*(1), 2–11.

Kocijancic, S., & Jamsek, J. (2004). Electronics courses for science and technology teachers. *International Journal of Engineering Education, 20*, 244–250.

Koehler, M., & Mishra, P. (2009). What is technological pedagogical content knowledge (TPACK)? *Contemporary Issues in Technology and Teacher Education, 9*(1), 60–70.

Kolb, D. A. (1984). *Experiential learning: Experience as the source of learning and development.* Prentice Hall.

Lewis, C. (2014). *Irresistible apps: Motivational design patterns for apps, games, and web-based communities.* Apress.

National Research Council. (1996). *National Education Standards.* National Academy Press.

National Science Teaching Association. (2005). *NSTA position statement.* Retrieved from: http://www.nsta.org/159&psid=16

Prajoko, S., Amin, M., Rohman, F., & Gipayana, M. (2016). The profile and the understanding of science process skills in Surakarta Open University students in science lab courses. In *Prosiding ICTTE FKIP UNS* (Vol. 1, pp. 980–985). ICTTE.

Ramesh, M. (2019). Virtual lab: A supplement for traditional lab to school students. *Research Review International Journal of Multidisciplinary, 4*, 434–436.

Sassi, E. (2000). Computer supported lab-work in physics education: Advantages and problems. In *International Conference on Physics Teacher Education beyond 2000.*

Sedumedi, T. D. T. (2017). Practical work activities as a method of assessing learning in chemistry teaching. *Eurasia Journal of Mathematics, Science and Technology Education, 13*(6), 1765–1784.

Shimba, M., Mahenge, M. P., & Sanga, C. A. (2017). Virtual labs versus hands-on labs for teaching and learning computer networking: A comparison study. *International Journal of Research Studies in Educational Technology, 6*(1), 43–58.

Skinner, B. F. (1968). Teaching science in high school—What is wrong? Scientists have not brought the methods of science to bear on the improvement of instruction. *Science, 159*(3816), 704–710.

Sullivan, S., Gnesdilow, D., Puntambekar, S., & Kim, J. S. (2017). Middle school students' learning of mechanics concepts through engagement in different sequences of physical and virtual experiments. *International Journal of Science Education, 39*(12), 1573–1600. https://doi.org/10.1080/09500693.2017.1341668

Tamir, P. (1991). Practical work in school science: An analysis of current practice. In B. E. Woolnough (Ed.), *Practical Science*. Open University Press.

Trisnaningsih, D. R., Parno, P., & Setiawan, A. M. (2021). The development of virtual laboratory-based STEM approach equipped feedback to improve critical thinking skills on acid-base concept. In *Advances in engineering research* (Vol. 209). Atlantis Press.

Trúchly, P., Medvecký, M., Podhradský, P., & El Mawas, N. (2019). STEM education supported by virtual laboratory incorporated in self-directed learning process. *Journal of Electrical Engineering, 70*(4), 332–344. http://iris.elf.stuba.sk/JEEEC/data/pdf/4_119-10.pdf

Woolnough, B. E., & Allsop, T. (1985). *Practical work in science*. Cambridge University Press.

Zacharia, Z. C., & de Jong, T. (2014). The effects on students' conceptual understanding of electric circuits of introducing virtual manipulatives within a physical manipulatives-oriented curriculum. *Cognition and Instruction, 32*(2), 101–158.

Zucca-Scott, L. (2010). Know thyself: The importance of humanism in education. *International Education, 40*(1), 4.

Further Reading

Abramov, V., Kugurakova, V., Rizvanov, A., et al. (2017). Virtual Biotechnological Lab Development. *BioNanoScience, 7*, 363–365. https://doi.org/10.1007/s12668-016-0368-9

Hernández-de-Menéndez, M., Vallejo Guevara, A., & Morales-Menendez, R. (2019). Virtual reality laboratories: A review of experiences. *International Journal on Interactive Design and Manufacturing, 13*, 947–966. https://doi.org/10.1007/s12008-019-00558-7

Miranda-Valenzuela, J. C., & Valenzuela-Ocaña, K. B. (2020). Using virtual labs to teach design and analysis of experiments. *International Journal on Interactive Design and Manufacturing, 14*, 1239–1252. https://doi.org/10.1007/s12008-020-00699-0

Shakeel M. C. Atchia has served the Mauritian Education system for the past 22 years and is currently an academic in the Science Education Department of the Mauritius Institute of Education (MIE). With a strong scientific background and a PhD in education, he has shown his interest in science education, STEM education, quality assurance, and comparative studies, through his publications, conference interventions and research work. He is currently the co-ordinator for Marine Science Education at MIE, the overall assistant Co-ordinator for the development of the National Curriculum Framework (NCF) of 'Technology Education' in Mauritius, the overall co-ordinator of a collaborative MIE-UKZN project on STEM education, and the international consultant for the Quantitative Analysis of the international project 'COVID-19 impacts on education systems', under the aegis of Stockholm University.

Anwar Rumjaun is an Associate Professor in the Science Education Department at the Mauritius Institute of Education (MIE). He is currently the Head of Research at MIE. His research interests are in science education, STEM, Environmental Education/ESD, including Climate Change. He is also the Curriculum Content Lead for the Master's curriculum for the module Research Methods and Dissertation for the SARUA. Currently a Senior Honorary Research Associate at the UCL Institute of Education, London, UK, he is also co-chairing the Scientific and Pedagogical Committee of the Climate Education Office, Paris, France.

Part II
Sustainable Development, Technology and Society

Chapter 10
Sustainable Development Goals and Science and Technology Education

Teresa J. Kennedy and Aletha R. Cherry

Abstract The Sustainable Development Goals (SDGs) are a collection of independent yet interconnected goals in support of the United Nations 2030 Agenda for Sustainable Development. The goals, created with the twenty-first century skills in mind, weave STEM (Science, Technology, Engineering and Mathematics) disciplines throughout and aim to end poverty, protect the health of our planet, and provide equitable educational opportunities to ensure that, by 2030, all members of civil society can enjoy prosperous and fulfilling lives. This chapter provides a historical perspective of the concept of sustainability, its relationship with global development, and the importance of developing a global STEM-literate workforce capable of responding to the worldwide challenges presented today and into the future. Implementation models of international non-governmental organizations (NGOs), as well as regional and national examples highlighting STEM programming related to formal, non-formal, and informal STEM educational settings promoting Education for Sustainable Development (ESD), are discussed.

Keywords Sustainability · Sustainable development · Sustainable Development Goals (SDGs) · Education for Sustainable Development (ESD) · Twenty-first century skills · STEM · STEM education · Formal · Non-formal and informal learning · Lifelong learning

T. J. Kennedy (✉)
The University of Texas, Tyler, TX, USA
e-mail: tkennedy@uttyler.edu

A. R. Cherry
Gradient Learning, Atlanta, GA, USA

© The Author(s), under exclusive license to Springer Nature Switzerland AG 2023
B. Akpan et al. (eds.), *Contemporary Issues in Science and Technology Education*, Contemporary Trends and Issues in Science Education 56, https://doi.org/10.1007/978-3-031-24259-5_10

Introduction

The *World Conservation Strategy*, a publication prepared in 1980 by the International Union for Conservation of Nature (ICUN), with advice from the United Nations Environment Program (UNEP) and the World Wildlife Fund (WWF), and in collaboration with the Food and Agriculture Organization of the United Nations (FAO) and the United Nations Educational, Scientific and Cultural Organization (UNESCO), provided the first formal introduction to the concept of global sustainability, with the aim of stimulating "a more focused approach to the management of living resources and to provide policy guidance on how this can be carried out" by government policy makers and their advisors, conservationists and those concerned with living resources, as well as by development practitioners such as agencies, industry and commerce, and trade unions (International Union for Conservation of Nature and Natural Resources, 1980, p. VI) . Six years later, an IUCN conference convened in Ottawa, Canada, to evaluate progress in implementing the plan, and concluded that the "emerging paradigm of sustainable development... seeks ... to respond to five broad requirements:

- Integration of conservation and development;
- Satisfaction of basic human needs;
- Achievement of equity and social justice;
- Provision for social self-determination and cultural diversity; and
- Maintenance of ecological integrity" (Purvis et al., 2019, p. 685).

"These challenges are so strongly interrelated that it is difficult, and indeed unhelpful, to arrange them in hierarchical or priority order. Each is both a goal in itself and a prerequisite to the achievement of the others" (Brooks, 1990, para. 16). The Strategy was later updated and renamed *Caring for the Earth: A Strategy for Sustainable Living* in 1991, expanding its reach to individuals and citizens' groups with the goal that global development would be sustainable. The message projected in this updated strategy was considered to be a visionary response to the 1987 report of the World Commission on Environment and Development entitled *Our Common Future*, often referred to as *The Brundtland Report*.

The *Brundtland Report*, written by Gro Harlem Brundtland, the first female prime minister of Norway and later Director General of the World Health Organization, suggested guidelines for best practices aimed at the preservation of biodiversity. A leader in the development of the renewed strategy, Brundtland stated, "humanity has the ability to make development sustainable to ensure that it meets the needs of the present without compromising the ability of future generations to meet their own needs" (World Commission on Environment and Development, 2011, p. 16). Her proclamation inspired the broad political concept of sustainable development, calling on citizens around the world to make philosophical and ethical considerations supporting intergenerational responsibilities aimed at addressing environmental concerns in the most effective manner possible.

Historical and Theoretical Background

Although the concept of sustainable development has taken on many interpretations since its inception, the core approach to development in a sustainable manner attempts to balance the different and often competing needs in support of environmental awareness, as well as broader foci such as the social and economic limitations that we face as a society, including "meeting the diverse needs of all people in existing and future communities, promoting personal wellbeing, social cohesion and inclusions, and creating equal opportunity" (Sustainable Development Commission, 2011, para. 5).

The United Nations (UN) Millennium Development Goals (MDGs) soon followed, setting eight goals, referred to as targets, to monitor the progress of efforts aimed at promoting gender equity, health, education, and environmental sustainability from 2005 to 2015. Figure 10.1 lists the eight UN MDGs.

However, monitoring the progress of each country's efforts related to the MDGs proved to be difficult, since countries inevitably faced different constraints when implementing targets that were set with global considerations in mind. As a result, some reports "moved away from comparing indicator levels across time and judging countries in terms of strict 2015 targets" (Hailu & Tsukada, 2011, p. 16). Nonetheless, the UN 2016 Annual Report noted that "Despite progress, the world failed to meet the MDG of universal primary education by 2015" (United Nations, 2016, p. 18).

During the implementation period of the MDGs, one of the most influential events setting the stage for the development of revised goals and/or targets was the 2012 United Nations Conference on Sustainable Development in Rio de Janeiro, where discussions concerning the MDGs began to expand at a significant pace. These discussions were the first steps toward creating a set of universal goals to help combat environmental, economic, and political challenges faced by all nations in the world, in a manner that could be made applicable within the context of each individual country.

Fig. 10.1 The United Nations Millennium Development Goals (MDGs). (Graphic used with permission by the Division of Science Policy and Capacity Building, UNESCO Natural Sciences Sector: https://www.un.org/millenniumgoals/)

The United Nations General Assembly worked beside more than 190 world leaders to create a worldwide plan aimed at achieving a more sustainable future. Their collaborative work resulted in 17 Sustainable Development Goals (SDGs), adopted in 2015 during the UN General Assembly as a call for all countries to work together toward a common vision. The SDGs are designed to end poverty and hunger, address worldwide health issues, and ensure racial and gender equity in all aspects of society, including equitable education and workforce opportunities for girls and women.

Shortly after the establishment of the SDGs, the 2030 Agenda for Sustainable Development was adopted by all United Nations Member States (in September 2015), demonstrating the multilateral consensus between governments, NGOs, businesses and education sectors around the world. Achievement of the SDGs is directly dependent upon the implementation of STEM (Science, Technology, Engineering and Mathematics) education on an international level, which in turn ensures the development of the essential skills needed for success in the twenty-first century workplace and beyond (problem-solving, critical thinking, creativity, improved communication skills, collaboration, perseverance, information and digital literacy, and entrepreneurial skills). See Chap. 4 for a detailed analysis of STEM education, and Chap. 20 for additional information on twenty-first century skills.

It is generally accepted that, by eliminating gender disparities, fostering an appreciation of cultural diversity, and promoting global citizenship, a global outcome leading to a better understanding of the contribution of cultures to sustainable development, and the collaborative actions needed to create a global workforce prepared to address current societal problems and those of the future, will emerge. In addition, the SDGs explicitly challenge institutions of higher education and businesses to promote creativity and innovation to address and solve present and future sustainable development challenges (see Chap. 19 for more detail on innovation and creativity). Figure 10.2 lists the 17 SDGs.

Fig. 10.2 Sustainable Development Goals. (Graphic used with permission by the Division of Science Policy and Capacity Building, UNESCO Natural Sciences Sector: https://en.unesco.org/sustainabledevelopmentgoals)

Sustainable Development Goal (SDG) 4, Quality Education, aims to "ensure inclusive and equitable quality education and promote lifelong learning opportunities for all" with the goal that "by 2030, all girls and boys will have access to quality early childhood development, care and pre-primary education so that they are ready for primary education" through SDG target 4.2 (United Nations, 2015).

While the tenets of science and technology education are naturally subsumed within this goal, it is also evident across the spectrum of goals. For example, students' education is enriched when they learn about, and take action against, poverty, pollution and equity—themes that are present throughout all the SDGs.

Formal and Informal STEM education and Its Relation to Sustainable Development

Academics generally agree that distinguishing between formal and informal educational contexts can be a difficult task. Therefore, it is important to clarify formal, non-formal and informal education in context and explain how these constructs fit into Education for Sustainable Development (ESD), a key instrument to achieve the SDGs.

The Organisation for Economic Cooperation and Development (OECD), an international forum consisting of the governments of 37 democracies representing market-based economies, recognized and defined formal, non-formal and informal education by structure and intended outcomes (2022). While formal education takes place within an educational institution with prescribed objectives, informal education often exists without such a plan, where learning is revealed through experiences. Over the last thirty plus years, museums, zoos, botanic gardens, horticultural gardens, field centers, science centers and geological sites, including those of industrial archaeology as well as cultural museums, have served as important centers for environmental education, adding to the growing interest in education for sustainable development (ESD) amongst science educators worldwide (Kennedy & Tunnicliffe, 2022; UNESCO, 2019).

Non-formal education is a midpoint, overlapping between an authoritative body governing educational journeys. "The wholistic nature of the learning process means that it operates at all levels of human society, from the individual to the group, to the organization, and to society as a whole" (Kolb & Kolb, 2009, p. 3). Kolb and Kolb contend that experiential learning theory encourages learning by doing, and spontaneous learning with no intended outcomes, further explaining that knowledge is gained from wholistic experiences, not just from a dependency of transfer from teacher to student. Learning is multilevel, gained from individuals and the environment. As noted by Jagušt et al. (2018), the most effective mode of learning is a combination of two types that are intentional, planned, and occur either inside or outside of the classroom.

The ideal delivery format for STEM education leads to an increase in curiosity, creativity, and action. French (2016) supports the idea that desirable learning experiences often occur when the student is engaged in something hands-on and of interest. "Engaging students in high quality STEM education requires programs to include rigorous curriculum, instruction, and assessment, integrate technology and engineering into the science and mathematics curriculum, and also promote scientific inquiry and the engineering design process" (Kennedy & Odell, 2014, p. 246). This concept of a personalized experience promotes student agency shaped around student choice, kinesthetic experiences and conversations that lead to activism, supporting a lifelong learning agenda for all that better matches the needs of the twenty-first century. In addition to a strong STEM education, 21st century skills, including cross-cultural skills, collaboration skills, critical thinking, and problem-solving, are needed to prepare students for success in the increasingly competitive global market (Kennedy & Sundberg, 2020, p. 479) as well as respond to the five requirements embedded in the *World Conservation Strategy* and its predecessor, *Caring for the Earth: A Strategy for Sustainable Living*. Consistent with the OECD mission to collaboratively develop policy standards aimed at supporting sustainable economic growth, all three educational contexts (formal, non-formal and informal) lead to the development of a strong global workforce.

STEM and Education for Sustainable Development (ESD)

SDG 4: Education identifies 10 targets to achieve the goal of inclusive and equitable quality education that can promote lifelong learning opportunities for all. By 2020, there was an expansion of scholarship opportunities available for tertiary students in developing countries in order to expand opportunities in higher education. The remaining targets address skills, access, and mindsets to provide educational opportunities for all women and men, girls and boys, persons with disabilities and indigenous peoples. There are many exemplary international programs that implement the tenets of STEM and ESD with fidelity. Examples of international STEM education programs that have demonstrated successful implementation of the SDG 4 targets follow:

Erasmus+

Erasmus+ is a program supported by the European Commission that promotes education, training, youth and sport throughout Europe. The 2021–2027 Erasmus priorities target social inclusion, green and digital transitions, and the importance of encouraging youth to participate in democratic life through educational mobility and co-operation opportunities for secondary and tertiary students, as well as those involved in vocational education and training. The Erasmus+ Project platform houses a comprehensive overview of all funded projects from 2021–2027 and highlights exemplary projects, practices and success stories. This innovative program promotes international networking and collaboration through its online system.

The GLOBE Program

The Global Learning and Observations to Benefit the Environment (GLOBE) program is a worldwide program that brings together students of all ages (primary, secondary and tertiary) with their teachers, scientists, and community members to promote science and learning about the environment, providing research investigations, and learning activities that involve making scientific measurements in five core fields: atmosphere, biosphere, hydrosphere, soil (pedosphere), and Earth as a system (GLOBE Program, 2020a). Each research area consists of validated measurement protocols and learning activities. GLOBE measurement protocols, developed collaboratively by international scientists and educators, allow participants to contribute standardized research-quality data. The U.S. National Aeronautics and Space Administration (NASA) provides and manages the programmatic infrastructure, coordinating and supporting GLOBE's international online community. Participants submit their local scientific observations to the GLOBE database, which currently contains over 200 million measurements from 127 countries around the world (GLOBE Program, 2020b). Research has shown that the GLOBE Program establishes a natural connection between the school and families, promoting cross-generational scientific understandings (Prieto & Kennedy, 2022). See Chap. 20 for more information about the GLOBE Program.

iEARN

The International Education and Research Network, known as iEARN, facilitates international collaborations with students in primary and secondary years and informal education groups (iEARN-USA, 2020). The projects feature tasks and ideas centered around STEM, social sciences, the Arts, literature and physical activity, while building essential learning skills, cultural understanding and friendships. Founded in 1988 by the Copen Family Foundation in the United States, it boasts a membership of 30,000 students in schools across over 100 countries and engages students and educators in more than 100 virtual collaborative projects. In spite of the COVID-19 pandemic, educators and students were able to increase confidence in cross-cultural skills and meet program goals for language, reading and mathematics proficiency.

Partners of the Americas

Founded in 1964 by American president John F. Kennedy, Partners of the Americas works to connect U.S. teachers and students with their counterparts in countries or regions of Latin America. The 94 chapters across more than 30 countries consist of young people, university-level students and adults in the fields of education and global citizenship, economic development and health, and child protection. The push for agricultural development and STEM collaborations through 100,000 people in the Americas and other programs enable participants to become change agents of sustainable communities and developed workforces.

The four examples listed above provide a small sample of exemplary international programs that promote STEM education in the context of ESD. These programs prepare students to graduate from their primary and secondary school experiences and continue to higher education STEM programs, or to enter the STEM workforce. There are currently more jobs available in STEM fields than qualified STEM applicants, a reality that reminds us of the importance of implementing quality STEM education programming, within the context of the SDGs, to close this gap.

Regional and National STEM education Initiatives and Research Findings

SDG integration ensures that complex global challenges can be addressed in a collaborative manner, and not in isolation. It also focuses on development that targets systems, not simply thematic sectors. Through collaborative system-based program development, SDG integration within a STEM education framework can address the root causes of an issue, as well as the inevitable ripple effects that impact economies, societies and natural ecosystems.

With this goal in mind, there are numerous exemplary regional and national initiatives implementing the interdisciplinary tenets of STEM and ESD with fidelity. These initiatives provide replicable models as well as innovative implementation ideas. A variety of approaches that increase STEM literacy to meet the targets of SDG 4, as well as the lessons learned and suggested recommendations for improvement, follow.

Regional and National STEM education Progress—Africa

"Education is often perceived in policy agendas as playing a transformative role in realizing sustainable development and the SDGs on the continent" (Tikly, 2019, p. 1), emphasizing the need for education to take on a transformative role, linked to processes of wider structural change across the economy, culture, and polity of countries throughout Africa. As an example, Kola (2020) described sustainable development initiatives in Nigeria, stating that the school serves as the foundation of all development initiatives in any twenty-first century nation. He argued that the best way to achieve sustainable development is to address three critical areas: education, governance and corruption. He further examined how to include technology during science instruction and learning in Nigeria, concluding that the lack of access to funds and modern technology slows the course of progress in teacher training, which has a direct impact on science teaching.

Science education must be able to convert science knowledge and skills developed in the classroom to relevant technology. Even then, while students can often successfully recite science concepts, they tend to lack the necessary skills for application. According to Kola et al. (2019), there is a small amount of those who can do science. However, a disconnect exists for students who only identify scientific concepts between their science education and sustainable development, negatively impacting the workforce and further reducing the merit of sustainable development. They believe that the link between science education and sustainable development is the teacher, concluding that "the attainment of excellence in high school Physics will collapse where there is no well-educated, strongly motivated, skilled, and well-supported teacher. Teachers at all levels of the educational system are paramount for the sustainable development of a nation" (Kola et al., 2019, p. 54). Tasked to cultivate students to transfer skills and apply their knowledge, teachers end up producing more students who can explain science or illustrate, but not *do* science. What students are lacking is linking their education to real-life, sustainable skills that are transferable in our dynamic society.

The literature offers solutions that highlight professional development and societal issues. For one, there should be an emphasis on Information and Communication Technology (ICT) to connect content, pedagogy and technology: the three core components of technology education. Additionally, teachers are encouraged to master the concepts so that students may experience authentic science learning. With real-life connections, active learning transfers to jobs that promote economical, ecological and human factors of sustainable development. According to Kola, currently many Nigerians see jobs exported out, and therefore recognize that they completed their education with no employable skills. The increasing number of under-qualified jobseekers trickle down to unskilled youth who have turned to violence in their communities, further supporting Kola's observation that "A science graduate who is rich in entrepreneurial skills will not be involved in any illegal business to degrade the environment" (2019, p. 35). Violence would inevitably be reduced with an increase in employable skillsets. Lastly, teacher training should double down on scientific literacy through TPACK (Technological Pedagogical and Content Knowledge) and its seven domains: Technology Knowledge, Content Knowledge, Pedagogical Knowledge, Pedagogical Content Knowledge, Technological Content Knowledge, Technological Pedagogical Knowledge, and Technological Pedagogical Content Knowledge. Additional recommendations encouraged teachers to allow for argumentation in the classroom, and to support students by allowing space for them to practice challenging classmates with claims and evidence. Kola postulates that student articulation and critical thinking skills will improve and break the current cycle through implementing the ESD reform initiatives described above. For more information about curriculum design as well as teaching strategies and pedagogical content knowledge, see Chaps. 5, 15, and 16.

Regional and National STEM education Progress—Asia and the Indo-Pacific

According to the Asia and the Pacific SDG Progress Report, the progress gap for achieving the SDGs has grown wider throughout the region due to "unsustainable development pathways coupled with an increase in the frequency and intensity of human-made crises and natural disasters" (Economic and Social Commission for Asia and the Pacific, 2022, p. 2), citing COVID-19 as the latest challenge that has pushed back the goals of the 2030 Agenda for Sustainable Development to at least 2065. Data gathered to date show that throughout the region, there is a critical need to reverse the negative trends impacting quality education as well as the inequality in access to education, citing a wide gender gap in regard to access to education and employment opportunities. "While the region has recorded a high enrollment rate, it is off track in ensuring students achieve [SDG] learning targets" (p. 36), especially those related to STEM.

For example, Amran et al. (2019) measured twenty-first century attitudes and awareness within six Indonesian senior high schools, examining skills in critical thinking, collaboration, communication, and creative thinking, which were evaluated alongside environmental awareness. Based on a questionnaire and self-assessment, critical thinking and creativity were evaluated as low due to the objectivist approach to teacher instruction and outcomes, while communication and collaboration were strong indicators for students who participated in extracurricular activities. Environmental awareness levels for the surveyed students highlight its importance, but not the application. School habits and culture, more than instructional practice, contributed to students' awareness. Based on the findings, they postulated that students could demonstrate higher competencies when teachers design lessons that consider the four aspects. "Science teachers are required…to teach the concept of science but need to exercise students' critical thinking skills" (Amran et al., 2019, p. 3). This study implicitly connects the value of student autonomy in science education to motivation and positive attitudes towards STEM subjects.

Regional and National STEM education Progress—Oceania

Oceania, consisting of 14 countries including Australasia (Australia and New Zealand), and spanning Melanesia, Micronesia and Polynesia, are confronted with specific and often acute challenges that affect their STEM development capacity and access to the global markets. These areas are also extremely susceptible to threats such as natural disasters and climate change, since many are small island developing states.

That said, sustainability is one of three national cross-curriculum priorities in Australia. Many schools and universities across the country incorporate the SDGs into their curricula and educational offerings. Questacon, the national science and technology center based in Canberra, provides teacher professional development as well as prepares students for STEM careers through traveling shows and demonstrations to all regions of the country, immersing students and their teachers in activities promoting creativity and other twenty-first century skills (United Nations, 2018, p. 41). In addition, Australian youth-led initiatives have been very successful, motivating students to volunteer to share their knowledge of STEM fields integrated with the SDGs at home and throughout the Oceania region. Examples include AIESEC (the International Association of Students in Economics and Business) in Australia, which sends young people to volunteer in countries throughout Asia as part of their *Youth for Global Goals* program, as well as the Australian Medical Students' Association's gender equity project focusing on SDG 5 and forming partnerships for action (United Nations, 2018, p. 10).

Regional and National STEM education Progress—Europe and Eurasia

The European region is heavily engaged in STEM education and its impact on achieving the SDGs, advocating for accelerated approaches to education and training through initiatives aimed at ensuring inclusive and equitable quality education and promoting lifelong learning opportunities for all (European Commission, 2015). Initiatives spreading information and communications technology (ICT) facilitate access to learning for all age groups, including young people and older adults, demonstrating the high level of importance placed on lifelong learning throughout the region. However, research revealed that, although the expenditure on education and share of tertiary education graduates in total population throughout Europe increased over the past two decades, the share of STEM program graduates in relation to the total tertiary graduates unfortunately declined (Bacovic et al., 2022).

The basic understanding that STEM education is the driver for sustainable development is key, since it raises awareness of critical issues and challenges. Current European initiatives under way to address this shortfall include integrated STEM activities to enhance policy coherence, improve the quality and equity of education and lifelong learning systems, develop education and training systems beyond their current reach at non-compulsory levels to provide lifelong learning and to reduce equity gaps in learning outcomes, ensure inclusive and quality education for refugees and migrants, strengthen the gender-education-health nexus, integrate partnership models, implement existing commitments and declarations, and enhance education-related data and monitoring systems (Regional United Nations Development Group for Europe and Central Asia, 2017, pp. 64–65).

A current European priority focuses on adapting and creating curricula to promote environmental awareness and training, with the overt goal of producing a green economy, while at the same time addressing the integrated goals associated with the 2030 Agenda. Several European projects organized by the International Council of Associations for Science Education (ICASE) have developed STEM competencies in line with these regional priorities. Two examples implemented from 2010–2014 are described below:

- ESTABLISH, the *European Science and Technology in Action: Building Links with Industry, Schools and Home* project, provided inquiry-based science education (IBSE) for European students aged 12–18 via online teaching and learning units; and
- PROFILES, the *Professional Reflection Oriented Focus on Inquiry-based Learning and Education through Science* project, shared IBSE amongst European science teachers through 'innovative learning environments' in 18 languages, providing long-term professional development with the aim of building science teacher self-efficacy to nurture future student success in science education.

Regional and National STEM education Progress—Latin America and the Caribbean

Three unique examples proving inspirational implementation models promoting STEM literacy and addressing the tenets of ESD are highlighted below from the Latin American and Caribbean region. The first demonstrates collaborative measures between institutions of higher education and a local youth center; the second describes the development of STEM learning societies; and the third highlights STEM's contributions to ecotourism in the region.

The University of São Paulo was highlighted by Santos et al. (2019) for its effective science practices involving an informal after-school STEM program for young people living in low socio-economic areas. Their Science Stand Project was developed by researchers at the University who partnered with students and professors from three public higher education institutions to implement the program within a youth center. Students were given choices of what to grow in their vertical herb garden. Playfulness, critical thinking and creativity were explicitly discussed and encouraged throughout the project. The facilitators introduced the students to the SDGs through *Smurfs for the SDGs*, a United Nations campaign that encouraged wellness and equality with videos, posters and interactive activities featuring the Smurfs and the celebrities who voiced the characters in a recent movie (United Nations, 2017). As the students developed their vertical gardens, they related to sustaining, first through their local lens, and then extended their thinking to larger environmental and human-initiated interaction. Photo walks provided perspective and opportunities for critical thinking. The project concluded with the young people sharing the benefits of vertical gardening through posters displayed throughout the community. Their conclusion survey revealed that this real-life experience allowed students the opportunity to relate more to their local environment and also increased their desire to share their knowledge beyond their neighborhoods.

Another notable program in the region was implemented in Colombia. When it was realized that the average period of study was 7.7 years, a decision had to be made, since the disadvantages of incomplete educational journeys can impact citizens for up to three generations. To address this problem, remote central regions were to be transformed into learning societies to distribute "knowledge in such a way that all societies have access to science and technology advances and that the most distant territories can be incorporated into global development, ensuring that they contribute to the necessary solutions" (Cujía et al., 2017, p. 187). Learning societies normalize science education across generations. Wealth, power, and knowledge, often by-products of higher education exposure, should be democratized. This concept, proposed by Joseph Stiglitz and supported by the UN, promotes higher standards of living as science and technology knowledge and skills develop. University researchers play a major role, as they bring inequities to light, and communicate possible research-based solutions while highlighting the responsibilities of all parties.

Government funding and support were vital for Colombia's vision of improved sustainable development and technology. In the case described above, regarding the city of Riohacha, La Guajira, the federal and local governments determined that ecotourism was the best avenue for the development of a region rich in biodiversity and cultural diversity. Their focus on ecotourism and innovation stimulated a positive impact on the city's infrastructure. By cultivating a shared learning experience of generating knowledge, the citizens have moved closer to sustainability and financial independence while remaining authentic to their community and culture.

Regional and National STEM education Progress—Near East and North Africa

Examples from the Near East and North Africa region focus on strategic course development as well as innovation and entrepreneurship as means of achieving the SDGs within a STEM framework. Research conducted in Iran showed that student awareness and critical thinking skills increased significantly through formal and informal educational experiences in sustainable development education at the university level (Pouratashi, 2021). Their findings inspired the revision of the academic content in existing courses, as well as the creation of specialized courses integrating socio-economic and environmental categories of sustainable development through interdisciplinary approaches to support twenty-first century skill development. They also encouraged their students to actively participate in professional associations and communicate with industry personnel to become better prepared for the STEM workforce.

Hemdan et al. (2022) also noted the importance of corporate partnerships in STEM education in the Near East region. For example, Rolls Royce's STEM Oman aims to promote participation in sustainable development through STEM-based competitions. These projects promote critical thinking and creativity and boast awards from all educational levels. Shell International's Nxplorer program focuses on solutions for SDGs that impact energy, water and food. BP Oman engages with students in cycle 1 classrooms (grades 1–4, ages 6–9 years) with Future Engineers to encourage social responsibility, and partners with the Engineering for Kids Association to bring Robotics Olympiad to schools. The country ensures sustainability by providing professional development for female teachers and discounts for students who desire to patent their ideas. Many similar regional opportunities were noted, including Kuwait's Marine Environment Competition for writing, which engages with several different corporate sponsors.

Regional and National STEM education Progress— North America

The North American innovations highlighted provide examples of STEM experiences in schools aimed at increasing student retention and preparation for the STEM workforce, and also include strategies to increase STEM activities in schools lacking resources, equipment, and materials.

Canada's approach to addressing SDG 4 is through efforts to reduce barriers to quality education, develop coaching and education programs targeted at specific groups, and implement indigenous education policy frameworks (Government of Canada, 2021). For example, in New Brunswick, the SDGs and climate action education have been embedded in its science curriculum for grades 3–10 (ages 8–15 years), and Ontario has engaged in curriculum modernization to better prepare students to enter the workforce. In addition, EcoSchools Canada offers an environmental certification program for students in kindergarten through to grade 12 (age 17 years). Their online platform provides access to over 45 SDG-connected environmental actions, facilitating schools across Canada's five distinct geographic regions to create their own environmental action plans.

Innovative efforts in the United States address various issues related to providing equitable STEM education opportunities for all students. The term *science on a shoestring,* and similar phrases relating to implementing STEM activities with an inadequate amount of funding to fully meet the needs of the intended outcome, acknowledge the financial gap that restricts educators from arranging the ideal environment for both child and adult learners. Limited budgets and resources result in educators adapting courses and spending more time gathering necessary items to contribute to learning. With the required support of volunteers from a college, worship center, and the local school district, Elrod et al. (1999) describe the success of a summer program for at-risk high school students. The program was chronically underfunded, and, at one time, the state ranked the lowest in the United States. However, the program saw an increase in grade-level skills up to 3 years, self-efficacy and excitement about school. Gartell describes the importance of thorough proposals for researchers, intense scheduling, and overcoming discouragement of declined grant proposals as solutions to address the problem of underfunded learning.

Conclusion

Integrated STEM education is at the core of the 2030 Agenda for Sustainable Development. Global development challenges will be addressed through implementation of the 17 Sustainable Development Goals (SDGs). Education for sustainable development requires a shift from 'teaching' to 'learning' and fully integrates all STEM disciplines. Combining formal, non-formal and informal education facilitates the advancement of ESD implementation, mobilizing all STEM courses taught through interdisciplinary connections to the SDGs.

While there are additional areas within education that have relevant connections to the SDGs' learner development, the disciplines subsumed by STEM described in this chapter hold particular significance to the SDGs. Students who are well-versed and literate in STEM fields tend to be creative, innovative and critical thinkers, capable of applying the information that they have learned to real-world problems, and they transition seamlessly into higher education careers in STEM fields.

The most recent reports at the time of this publication saw developments in some areas, but also overall setbacks due to the COVID-19 pandemic. As formal education reopens its doors, the lack of drinking water, basic sanitation and computers with Internet access becomes more apparent. These school necessities are improving as students return, but the inequities do not disappear. Additionally, there is a gap in psychological and emotional support due to trauma experienced as a result of the virus's impacts.

There are many who have been able to benefit from distance learning. However, for those students without the option of distance learning, there is concern that those from early years through college will not return after a significant time away (United Nations, 2022). As the first SDG report indicated, success in education was contingent upon the educational level of heads of households. SDG 10 (Reduced Inequalities) continues to be of concern, directly impacting the plan of SDG 4's Quality Education. Experiential STEM education enables the next generation of innovators and policymakers, who are prepared to examine problems and create plans to solve them, and ultimately work toward a sustainable future.

As a final note, we must never forget that ESD prepares all student to enter the STEM workforce. According to Irina Bokova, UNESCO Director General, both education and gender equality play strong roles in the 2030 Agenda for Sustainable Development, as distinct goals, but also as catalysts to the success of all the SDGs. In addition to targeting all students to engage in STEM disciplines, we need to "understand and target the particular obstacles that keep female students away from STEM. We need to stimulate interest from the earliest years, to combat stereotypes, to train teachers to encourage girls to pursue STEM careers, to develop curricula that are gender-sensitive, to mentor girls and young women and change mindsets" (UNESCO, 2017, p. 5).

Summary

This chapter covered historical and contemporary issues found within science and technology education and provided links to the United Nations Sustainable Development Goals (SDGs) through learning experiences from around the world. A reoccurring theme centered on the importance of focusing on multiple SDGs, through interdisciplinary implementation of STEM disciplines, to fully address SDG 4: Quality Education. Formal, non-formal and informal educational environments provide valuable opportunities for STEM integration within the SDG framework, making science and technology relevant to students.

Transformations that lead to quality science and technology education must include partnerships, such as through institutions of higher education, corporations, and government entities, and should center around quality learning experiences. STEM disciplines promote the twenty-first century skills needed to be strong advocates and champions for the SDGs, preparing students to enter the workforce and to engage with governments, local authorities, and institutions.

Recommended Resources

For Institutions of Higher Education:
- UNESCO Science Report https://unesdoc.unesco.org/ark:/48223/pf0000235406
- UNESCO Engineering Report https://en.unesco.org/reports/engineering

For Primary and Secondary Teachers
- Resources for Educators https://en.unesco.org/themes/education/sdgs/material
- Teach the SDGs http://www.teachsdgs.org/resources.html
- UN Supported Publications (Science and Goal 4) https://sdgs.un.org/publications?field_review_year_value=&field_publisher_value=&tid=1193&goals=4
- Achieving the SDGs with Children https://www.unicef.org/sdgs/how-achieve-sdgs-for-with-children
- Smithsonian Science for Global Goals https://ssec.si.edu/global-goals
- SDG Unit Lesson in Spanish https://www.unescoetxea.org/dokumentuak/Unidad-didactica-ODS-completo.pdf

- Indicator Guide for SDG 4 in Spanish https://issuu.com/educationinternational/docs/2017_sdgs_toolkit_esp_v1.1
- Teacher Training and the SDGs in Portuguese https://portaleventos.uffs.edu.br/index.php/EIE/article/download/15144/9930/

For Students

- SDG Student Resources https://www.un.org/sustainabledevelopment/student-resources/
- Images of the Smurfs with SDGs https://www.speakactchange.org/2018/11/14/sdgs-small-smurfs-big-goals/

For Teacher and Student International Collaboration

- Erasmus+: https://erasmus-plus.ec.europa.eu/contacts/national-agencies
- GLOBE: www.globe.gov
- iEARN: https://us.iearn.org/
- Partners of the Americas: https://www.partners.net/

References

Amran, A., Perkasa, M., Satriawan, M., Jasin, I., & Irwansyah, M. (2019). Assessing students 21st century attitude and environmental awareness: Promoting education for sustainable development through science education. *Journal of Physics, Conference Series, 1157,* 022025. https://iopscience.iop.org/article/10.1088/1742-6596/1157/2/022025

Bacovic, M., Andrijasevic, Z., & Pejovic, B. (2022). STEM education and growth in Europe. *Journal of Knowledge Economy, 13,* 2348–2371. https://doi.org/10.1007/s13132-021-00817-7

Brooks, D. (1990, October). Beyond catch phrases: What does sustainable development really mean? *IDRC Reports.* https://idl-bnc-idrc.dspacedirect.org/bitstream/handle/10625/22894/108930.pdf?sequence=1

Economic and Social Commission for Asia and the Pacific. (2022). *Asia and the Pacific SDG Progress Report 2022: Widening disparities amid COVID-19.* ESCAP 75. https://www.unescap.org/sites/default/d8files/knowledge-products/ESCAP-2022-FG_SDG-Progress-Report.pdf

Cujía, E., Pérez, S., & Maestre, D. (2017). Ecoturismo, educación, ciencia y tecnología, factores de desarrollo sustentable: caso La Guajira, Colombia. *Revista Educación y Humanismo, 19*(32), 174–189. https://doi.org/10.17081/eduhum.19.32.2540

Elrod, G., Blackbourn, J., Mann, M., & Thomas, C. (1999). Shoestring partnerships: A pilot project on community involvement in serving at-risk youth. *National Forum of Applied Educational Research Journal, 12E,* 9–16. Overlook Press.

European Commission. (2015). *Science education for responsible citizenship.* Report to the European Commission of the Expert Group on Science Education. https://www.researchgate.net/publication/280831573_Science_Education_for_Responsible_Citizenship

French, D. (2016). Student agency and personalized learning. *Center for Collaborative Education.* http://cce.org/thought-leadership/blog/post/student-agency-and-personalized-learning

GLOBE Program. (2020a). *About GLOBE.* https://www.globe.gov/about/overview

GLOBE Program. (2020b). *GLOBE impact around the world.* https://www.globe.gov/about/impact-and-metrics

Government of Canada. (2021). *Taking action together – Canada's 2021 Annual Report of the 2030 agenda and the sustainable development goals.* https://www.canada.ca/en/employment-social-development/programs/agenda-2030/taking-action-together.html#sdg4

Hailu, D., & Tsukada, R. (2011, February). *Achieving the Millennium Development Goals: A measure of progress*. United Nations Development Programme (UNDP). https://www.undp. org/publications/achieving-millennium-development-goals-measure-progress?utm_source= EN&utm_medium=GSR&utm_content=US_UNDP_PaidSearch_Brand_English&utm_ campaign=CENTRAL&c_src=CENTRAL&c_src2=GSR&gclid=Cj0K CQjw852XBhC6ARIsAJsFPN0cBNoG8KnYIwBMP8fmwOq7j_tAFlL RmFyqh6CFX1EhsLXEM5BdUX4aAtYpEALw_wcB

Hemdan, A. H., Ambusaidi, A., & Al-Kharusi, T. (2022). Gifted education in Oman: Analyses from a learning-resource perspective. *Cogent Education, 9*(1), 2064410. https://doi.org/ 10.1080/2331186X.2022.2064410

iEARN-USA. (2020). *iEARN-USA 2020 Annual Report*. International Education and Resource Network. https://us.iearn.org/assets/imgs/docs/iEARN-USA-Annual-Report-2020.pdf

International Union for Conservation of Nature and Natural Resources. (1980). *World conservation strategy: Living resource conservation for sustainable development*. IUCN-UNEP-WWF. https://portals.iucn.org/library/efiles/documents/wcs-004.pdf

Jagušt, T., Botički, I., & So, H. (2018). A review of research on bridging the gap between formal and informal learning with technology in primary school contexts. *Journal of Computer Assisted Learning, 34*(4), 417–428.

Kennedy, T. J. & Odell, M. R. L. (2014). Engaging students in STEM education. *Science Education International, 25*(3), 246–258. https://eric.ed.gov/?id=EJ1044508

Kennedy, T. J., & Sundberg, C. W. (2020). 21st Century Skills. In B. Akpan & T. J. Kennedy (Eds.), *Science Education in Theory and Practice: An Introductory Guide to Learning Theory* (Chapter 32, pp. 479–496). Switzerland: Springer Nature. https://doi.org/10.1007/978-3-030-43620-9

Kennedy, T. J., & Tunnicliffe, S. D. (2022). Introduction: The role of play and STEM in the early years. In S. D. Tunnicliffe & T. J. Kennedy (Eds.), *Play and STEM Education in the Early Years: International Policies and Practices* (Chapter 1, pp. 3–37). Switzerland: Springer Nature. https://link.springer.com/chapter/10.1007/978-3-030-99830-1_1

Kola, A. J. (2020). Sustainable Development challenges in Nigeria: The role of science Education. *Sumerianz Journal of Education, Linguistics and Literature, 3*(4), 32–37.

Kola, A. J., Gama, N. N., & Olu, A. M. (2019). The trajectories of science education in Nigeria and its challenge to sustainable development. *Cross-Currents: An International Peer-Reviewed Journal on Humanities and Social Sciences, 5*(3), 53–61. https://www.researchgate.net/ profile/Aina-Kola-2/publication/332866771_The_Trajectories_of_Science_Education_in_ Nigeria_and_its_Challenge_to_Sustainable_Development/links/5ccecd29a6fdccc9dd8db176/ The-Trajectories-of-Science-Education-in-Nigeria-and-its-Challenge-to-Sustainable-Development.pdf

Kolb, A. Y., & Kolb, D. A. (2009). Experiential learning theory: A dynamic, holistic approach to management learning, education and development. In S. J. Armstrong & C. V. Fukami (Eds.), *The sage handbook of management learning, education, and development* (pp. 42–68). Sage Publications Ltd.. https://sk.sagepub.com/reference/hdbk_mgmtlearning/n3.xml

OECD. (2022). *Recognition of non-formal and informal learning – Home* [Web page]. Retrieved from: https://www.oecd.org/education/skills-beyond-school/recognitionofnon-formalandinformallearning-home.htm. Accessed 21.07.22.

Pouratashi, M. (2021). The influence of formal and informal education on students sustainable development skills, a study in Iran. *Zagreb International Review of Economics & Business, 24*(2), 25–35. https://sciendo.com/article/10.2478/zireb-2021-0009

Prieto, A. B., & Kennedy, T. J. (2022). GLOBE, STEM and Citizen Science in the Early Years in Argentina. In S. D. Tunnicliffe & T. J. Kennedy (Eds.), *Play and STEM Education in the Early Years: International Policies and Practices* (Chapter 19, pp. 383–423). Switzerland: Springer Nature. https://link.springer.com/book/10.1007/978-3-030-99830-1

Purvis, B., Mao, Y., & Robinson, D. (2019). Three pillars of sustainability: In search of conceptual origins. *Sustainability Science, 14*, 681–695. https://doi.org/10.1007/s11625-018-0627-5

Regional United Nations Development Group for Europe and Central Asia. (2017). Building more inclusive, sustainable and prosperous societies in Europe and Central Asia. *Regional Advocacy Paper*. https://unece.org/sites/default/files/2020-12/ECA_Regional_Advocacy_ Paper_2017_0.pdf

Santos, W., Singh, D., da Cruz, L., Piassi, L., & Reis, G. (2019). Vertical gardens: Sustainability, youth participation, and the promotion of change in a socio-economically vulnerable community in Brazil. *Education Sciences, 9*(161), 1–14. https://doi.org/10.3390/educsci9030161

Sustainable Development Commission. (2011). *What is sustainable development?* https://www.sd-commission.org.uk/pages/what-is-sustainable-development.html

Tikly, L. (2019). Education for sustainable development in Africa: A critique of regional agendas. *Asia Pacific Education Review, 20*, 223–237. https://link.springer.com/content/pdf/10.1007/s12564-019-09600-5.pdf

UNESCO. (2017). *Cracking the code: Girls' and women's education in science, technology, engineering and mathematics (STEM).* Education 2030. https://unesdoc.unesco.org/ark:/48223/pf0000253479

UNESCO. (2019). *What is education for sustainable development?* [Web page]. United Nations Educational, Scientific and Cultural Organization. https://en.unesco.org/themes/education-sustainable-development/what-is-esd

United Nations. (2011). *Our common future: Report of the World Commission on Environment and Development.* https://sustainabledevelopment.un.org/content/documents/5987our-common-future.pdf

United Nations. (2015, October 15). *Resolution adopted by the General Assembly on 25* September 2015. Seventieth session, Agenda items 16 and 116. https://www.un.org/ga/search/view_doc.asp?symbol=A/RES/70/1&Lang=E

United Nations. (2016). *The sustainable development goals report.* United Nations Department of Economic Affairs. https://unstats.un.org/sdgs/report/2016/The%20Sustainable%20Development%20Goals%20Report%202016.pdf

United Nations. (2017). *Ahead of International Day, UN and Smurfs team up to promote happiness and sustainable development.* [Web page]. United Nations. https://www.un.org/sustainabledevelopment/blog/2017/03/ahead-of-international-day-un-and-smurfs-team-up-to-promote-happiness-and-sustainable-development/

United Nations. (2018). *Report on the implementation of the sustainable development goals.* United Nations High-Level Political Forum on Sustainable Development. Australian Government. https://www.dfat.gov.au/sites/default/files/sdg-voluntary-national-review.pdf

United Nations. (2022). *The sustainable development goals report 2022.* United Nations Department of Economic Affairs. https://unstats.un.org/sdgs/report/2022/The-Sustainable-Development-Goals-Report-2022.pdf

Teresa J. Kennedy is a Professor of International STEM and Bilingual/ELL Education in the College of Education and Psychology at the University of Texas at Tyler. She holds appointments in the School of Education and the College of Engineering. She is a Past President of the International Council of Associations for Science Education, and currently serves on the ICASE Executive Committee, is a member of the UNESCO NGO Liaison Committee, and serves on the NSTA International Advisory Board. Her research interests include international comparative education, gender and equity issues in STEM, and content-based second language teaching and learning focused on STEM disciplines. Dr. Kennedy holds a PhD in Curriculum and Instruction from the University of Idaho.

Aletha R. Cherry is the Elementary Lead at Gradient Learning. She is an Education Ambassador for the United States of America and the Federative Republic of Brazil. She currently sits as Vice President for the Georgia Chapter of Partners of the Americas and serves on the NSTA International Advisory Board. Research interests include whole child practices, intersectional experiences of Afro-descendant women in the Americas, and international STEM education collaborations. Dr. Cherry holds a PhD in Curriculum and Instruction from Texas Tech University.

Chapter 11
In the Beginning: Interpreting Everyday Science

Sue Dale Tunnicliffe

Abstract This chapter explores ways in which children begin to learn science, in its different manifestations, from their earliest days. It recognises that science is inextricably intertwined with other STEM subjects. Above all, it stresses the importance of hearing the voice of the child, not ignoring it and their own understandings, listening to the child, not just telling them things, scaffolding their scientific literacy development and acquisition of science capital. The importance of out-of-school learning is recognised, including in museums of the widest genre. In the beginning, the youngest children observe and make sense of their world, including the science action in their lives, leading to science understanding upon which the theory can be built if formal education is available. The earliest interactions with everyday science are experiences, but are also the foundation for future understanding and learning. These are the foundation upon which science teachers can build to assist the learner in school in constructing the knowledge of the curriculum.

Keywords Sciences · Early years · child's voice · Formal school · Informal sites · Play · Progression

Introduction

Experiencing science in action begins from birth. The youngest of children observe and make sense of their world. They actively engage in physical science actions and observe the effects of Earth science, biological organisms and phenomena, experiencing the biology in action in themselves as living beings. Such learning is observational and experiential. Adults, the media and venues such as museums,

S. D. Tunnicliffe (✉)
Reader in Science Education, University College London Institute of Education, London, UK
e-mail: s.tunnicliffe@ucl.ac.uk

zoos, gardens and, indeed, the everyday environment are sources of information. Informal science is now a recognised part of learning about science for all ages. Such experiential learning is the laying of the foundations of their scientific literacy and understanding. It is different from formal school science. Should teachers seek to develop the child's observations and experiences in school, rather than teach the conventional science curriculum as most countries prescribe? UNESCO has suggested that schools should build on the science most relevant to everyday life, such as human biology and health, and climate science. With the human issues confronting our planet, isn't such understanding essential for sustainability and survival? Abrahams (2021) has questioned whether current science education is really relevant in today's world, with so many adults in the UK and globally not understanding about disease, transmission, immunity and prevention, as was shown during the COVID-19 pandemic. In essence, the starting point for learning science is play. However, the recognition and study of *Early Childhood Education* has been largely neglected when researching into the opportunities for learning science and associated subjects (such as STEM), but is beginning to be recognised as critically important (Milford & Tippett, 2015).

What Is science Learning?

Where does this occur? Should science be taught from the earliest stages of formal education up to leaving school? This was the question posed by Eschach and Fried (2005). The concept of science as purely an artificially conceived set of subjects in school and learning is unrealistic. Science is observed, experienced and learnt in different places: formally in school during the statutory years of education; in out-of-school locations designed to provide learning opportunities; science and natural history museums; science centres and zoos. Science in action is experienced and observed in leisure time through various forms of media, including the Internet and television, and through books as well as certain leisure activities with a science content. Science is evidentially-based. Children, either in pre-formal school settings such as kindergarten or nurseries, or in their play at home and in the community, do collect various aspects of evidence as they interpret observations and outcomes. Much early science is also identifiable as early maths and engineering.

Monteira and Jiménez-Aleixandre (2015) worked with Spanish children aged 5–6 years in a kindergarten class engaged in science activities and found that these children could distinguish between evidence from their own observations, in this case, on snails, as opposed to the empirical evidence from planned investigations with the teacher. Some formal education practitioners value play activities, recognising that, for these early learners, science ideas and practice are integrated with familiar early years strategies such as storytelling. Vartiainen and Kumpulainen (2020) identified aspects of scientific play during activities that were enquiry-based with a pre-school group in Finland. Carruthers and Worthington (2006) wrote about the beginning of mathematical understanding. So-called children's science (Osborne et al., 1983) takes place during children's engagement with the world in which they live and to which they give meaning through their actions, experiences, current knowledge and language. This sense-making takes place through the search for similarities and differences, through the organisation of events and phenomena, and through the observation of their environment. In this way, children collect data in a certain way, look for explanations, form models and make predictions. The tremendously compelling curiosity of these young children, before formal education, and their implicit desire to understand their living and inanimate environment, are what drive children to explore their environment actively, curiously and generously, albeit more unsystematically and less stringently than scientists.

Science as a concept has constituent parts. We do the whole concept harm by treating it as a uniform entity. The most important category to recognise at the beginning is the biological domain, because we, practitioners and learners in this specific case, are biological organisms who, in our physiology and anatomy, utilise principles of the physical domain. We experience biology in action, which can provide a personal point of view when encountering and observing biological organisms and systems. As living beings, we observe natural phenomena that have shaped our species. Earth science has formed the natural environment, providing biomes and climates in which organisms live, as well as the weather and other climate functions that affect us. We also encounter and co-exist with other inhabitants of this world, including organisms that are beneficial to us, others that are not and also some seeking to live in harmony with us. It is important for educators to understand both the ways in which beginning learners acquire information and their ideas about their everyday world, and also that children come to our lessons not as *tabula rasa,* but as *children scientists.*

Formal and Informal: science Learning Opportunities

Finding out and learning about science might occur in three distinct locations: (1) homes; (2) other venues, purpose-built for the dissemination of information, including science museums, cultural museums, botanical gardens, zoos and aquaria; and (3) in formal education settings that are implementing governmental policies and practice.

Children spend far less time in formal learning establishments than they do in the informal ones. Hence, informal learning is an important part of developing science capital and complementing school learning. Formal learning occurs within school, including as part of the curriculum, and also outside the classroom through activities in the school grounds. Formal learning also occurs in other venues, such as in museums. Informal learning takes place beyond the auspices of the school, in leisure time in the learner's community or at home. However, in out-of-school venues, educators consider whether or not the learner has voluntarily visited the site or whether they are 'conscripts' led by educators (Smith et al., 1998). Children taken to a museum by parents or carers during out-of-school time, or, in the case of pre-formal school learners, in playgroup or childcare situations, *are* in fact conscripts too, because they are taken by others, although, on occasion a young child will ask to be taken, often for a repeat visit to a pond or zoo, for example. Children's playgrounds are a site of experiential STEM learning.

Ros Driver published her seminal book, *The Pupil as Scientist?* in 1983 (Driver, 1983), which revolutionised the *thinking about their thinking* of secondary (11–18 years) science teachers in the UK. Driver pointed out that secondary pupils had ideas about science learning and that the pupil (of secondary age) was a scientist. These pupils came to formal school science lessons with relevant experiences and their understanding, albeit not always in agreement with the interpretations of scientists of the same phenomena. In the last third of the twentieth century, changes happened in science education. A paradigm shift recognised that pre-secondary children in school were also capable of learning science if it was taught in appropriate ways. ICASE started a pre-secondary science education section in 1988 at their Canberra meeting. It is during the first quarter of the twenty-first century that acknowledgment and understanding that pre-school children also experience and display intuitive science has emerged. Two science educators in the UK founded the *Journal of Emergent Science* in 2012 to cover the reporting of relevant research in science learning for this age group and their practitioners. A similar move occurred through the National Science Teachers Association (NSTA). Alison Gopnik published her research in a form accessible to practitioners through her books aimed at a general readership, which showed that the youngest of children, through their activities, demonstrate the attributes of science investigations (Gopnik, 2012). School science is but one aspect of science in the world. Remarkably, even practising scientists fail to recognise science in action in their lives, considering science to be that which is restricted to laboratories or field research. Towards the end of the first quarter of the twenty-first century, the realisation that the youngest children are also actively experiencing STEM through play and everyday tasks is leading to innovatory practices, policies and research.

The term *'science'* is frequently used to cover STEM areas. During the first two decades of the twenty-first century, it has been recognised that these areas are often inextricably linked. The rigid division between science and the Arts that emerged in the second part of the nineteenth century is now being broken down gradually. In the 1980s, the UK Government introduced a national curriculum, which included science and design and technology, to be taught from the statutory start of school at 5 years of age, and the *Early Years* framework for children under 5 in pre-schools, which included *Understanding about the World*. Increasingly in UK primary school (5–11 years), science activities embrace what is often now referred to as basic engineering, and encapsulate some of mathematics concepts as well as some mathematics skills such as measuring the collection of data and contrasting tables and graphs. Early formal schooling focuses on developing literacy in language, speaking, listening, reading and writing, as well as basic numeracy and other mathematical areas, but does not recognise and link these STEM areas as shared experiences.

Moreover, the traditional content of science teaching is increasingly being challenged. The surge of a media environment and issues of economics and climate change is challenging further the content of science curricula. Such observations have led to the recognition of an increasing involvement of citizens through, for example, social media and, in some cases, citizen science projects. An understanding is emerging that science relevant to the people is involving them in active participation, unlike school science formal teaching. Thus, a realisation is developing that science in the community and media is also a tool to foster an informed citizenship, who can participate in an informed way in decisions. Increasingly, through an understanding of science in the community outside the formal school set-up, educators are challenging the content of relevant science teaching for most pupils. There are fears that the level of science literacy in school leavers is inadequate to grasp the issues of the present times, and schools should perhaps focus on the a more targeted and functional scientific literacy to be developed and evaluated in schools. These most recent thoughts were prompted by the COVID-19 pandemic, where the understanding of viruses and immunisations were found to be largely lacking.

A constraining influence on education policies and practice in many countries is the effect of the neoliberalism polices of many governments (Roberts-Holmes & Moss, 2021). A business model is applied, and accountability measured by test results of the children, publications of league tables of school examination results and inspection reports. In England, more and more schools are being taken under the control of boards made up of business people looking for financial results as they establish Academies, often in groups, which receive funding directly from the government and over which the Local Authorities, hence the people through their elected members, have no control or input. The child is the only player in this scenario who is not consulted, but is the object to be educated to satisfy the demands of the system in place.

In the Beginning

A Child's First Encounters with Science

Increasingly, practitioners and others are recognising the importance of understanding and facilitating the first experiences of the youngest children with science, in its widest sense, in their everyday. However, we are part of biology, as are biological organisms. As children develop, they tend to interpret anthropomorphically much of the biology that they encounter, transferring what they understand as their own needs, rather as Maslow's hierarchy found. Such anthropomorphic behaviour is exhibited by many adults in their interpretation of animal behaviour, particularly with domestic pets and when talking about zoo animals, particularly the primates. There are three dimensions to young children becoming aware and beginning to interpret everyday biology, but this awareness is not uniform. Tunnicliffe (2020) identified three dimensions, namely: *Time, Observations* and *Systems.*

There are no knowledge boundaries for these emergent learners making sense of their world, so the divisions of formal school curricula in science are not apparent to them. Progression in understanding meets no barriers. The majority of toys with which Western children play require pushes and pulls in the early stages. It is not only toys with which the early learners interact, but also everyday objects such as doors, which, if open and the child is able to crawl, can be pushed open or closed, so using and experiencing a force in action. If they have a toy with wheels, they learn about pulling and tension. Pushing a wheeled vehicle by hand over different surfaces offers the child an experience of the effects of friction. These are but a few examples of physical science in action.

Physical science is the predominate domain in action play with toys, which is optimally in a free choice setting. As children allow objects to drop to the floor, they observe the effects of gravity, but also, having the toy returned to the child as they sit in a pram or highchair allows the child the opportunity to begin to collect data and develop the scientific process. Earth science becomes important in the youngest child's experiences. The change between day and night, weather patterns and their effects on the toddler (a need for appropriate clothing and what activities are allowed) all provide experiential learning. Although there are some hands-on Earth science experiences, such as sifting silt, collecting small stones and pebbles, playing with mud, walking on different surfaces and elevations, noticing landscapes, much is observational, as is biology in their world other than first-hand experiences of themselves in life processes. These domains are less frequently apparent in early years settings where the physical domain is predominant. We practitioners have to consider carefully how Earth science and biology can realistically and constructively become part of a child's experiences from birth. As soon as children meet objects, they begin to experience shapes, mass, size, colour and different materials. In their explorations and early play, for example with blocks, they are experiencing fundamental maths concepts and, most importantly, spatial thinking (Pollman, 2010). Also, in the structures that they build with toys and also in adventure areas, they are intuitively using engineering principles. When exploring their environment or playing with toys, children use a variety of approaches that show similarities to those used by scientists, such as observation, classification or data collection (Gopnik, 2012). Without science-aware practitioners, so much of children's science learning associated with play, free choice or in learning environments is not recognised. The experience of science in action is part of play (Tunnicliffe & Gkouskou, 2019).

Play

Play is the universal activity of the emergent learner in some mammals, including humans. The types of play, location and toys used vary. Adults have an influence on a child's play in some ways. Researchers maintain that play in ancestral communities and for some present-day indigenous people is an apprenticeship for adulthood and that these children play with adult items in miniature (Riede et al., 2018). Young children in many societies also re-enact adults' actions and behaviours using miniature objects. Play is associated with the children initiating it themselves, not with adults instructing. The pre-school years are often popularly regarded as 'just playing', not as real learning, but many are beginning to recognise that play in the pre-school years is a vital component in a child's development and involves much STEM learning though experiences. Play is play. We can identify in play elements not only of the STEM subjects, but also signs of socialisation, problem-solving and physical development and progression. Number, measurement, space and time are important math concepts. As pre-school children learn through play and the world around them, they are experiencing STEM basics. It is the adults who become concerned about what the child is learning, as they give the actions labels. There is commonality in early interactions, whether they are labelled maths, science or engineering, such as making collections, or constructing towers that are stable.

The child's idea of number is associated with length and space occupied, not by individual items present. Children begin to understand 'how much?', 'how many?'. Block play helps the early learner learn about balance, measurements, space, shapes and number. The role of an adult is important in using mathematics words in everyday life, counting how many of something, for example, the blocks that they have, the number of things in a particular colour, how many items they have collected in their basket as they finish rushing round a room pushing the doll's pushchair. Hearing adults say the words when cooking, for example, measuring out ingredients or collecting slices of bread to make toast, is important, as well as how many mugs to put out for their mid-morning drink at playgroup.

Children enjoy sorting, matching and counting everyday things such as bottle tops and counters, as well as putting items into a container and taking them out again. They make sets according to their own system, such as putting all toy cars in one place and all toy animals in another. Measuring is very important when involved in play. Water play is very much a science and maths activity, pouring from one-sized container to another, and then into different-shaped containers but of the same volume. Children develop maths concepts gradually and through experiences in play. As children learn about number, measurement, space and time, they are learning about the relationships between objects, and about their own relationships to objects and to the world around them. Using numbers comes naturally to children. They enjoy counting how many items they are holding, or how many children, or family, in a home situation. Children who practise counting lots of things have an easier time learning about numbers. They also enjoy making collections of stones and pebbles, toy cars, sticks or leaves, and often such collections can be used to develop ideas of science, particularly for botany and Earth science, as well as using descriptive words and counting, adding, subtracting, multiplying and dividing up the components of such collections. Young children's ideas about spatial relationships are very much based on how they look to them. From about 1 year old, a child realises that even though they can no longer see an object, it still exists. Three-year-olds, with their understanding of what they can see, learn directions when they want something that they cannot see, such as their coat, or a toy.

The STEM play Cycle, Free Choice and Progression

Stages in a STEM Play Session

Play is not uniform. Firstly, the term 'play' is a superordinate category covering the activities of young children observing and interacting with natural phenomena, such as the weather, or constructed phenomena, such as everyday items or toys, created specifically for children by adults, and frequently not used in the way in which the adult envisages.

Secondly, play is progressive in terms of understanding, capabilities and skills. Experiences are honed and developed. Play is not the same in the earliest years as it is in adulthood. Thirdly, play interactions depend on the interests of individual children.

The initial but critical stage in a child 'playing' is that their interest is 'caught' and subsequently maintained, so that they enter into an interaction with the phenomenon. This is the basis of the person-object theory of interest (POI), where the initial interest elicited by the first encounter is caught. The interest represents a particular interest, particular to that child, between the child and the phenomenon, usually an object. It includes initial attention to the item and interacting in a task, requiring maintained curiosity and engagement. Whilst widely applied in museum work with visitors at exhibits, this theory is pertinent to play. Much 'science'-focused play and, indeed, activities in playgroups and nurseries with constructed equipment, take a physical science manifestation involving pushes and pulls, essentially using sources that are powered by the energy from the child's play and social action. Hence, I developed the play sequence or cycle. My STEM play cycle has distinct identifiable stages that can be identified when observing children at play and is being used by CASTME (the Commonwealth Association of Science, Technology and Mathematics Educators). There are variations depending on the science inherent in the experience and interest object or phenomenon. The key starting point is that the children are intrigued and explore the artefact or phenomenon that they encounter.

If a child is observed playing, the stages though which they pass can be identified. A child may lose interest before reaching the end of the cycle, they may return later having had a 'think', or they may not return. However, using the sequence provides an insight into a child's thinking and intuitive actions. The interactions follow a *STEM Play Cycle* similar to the one that occurs in enquiry science investigation in school (Fig. 11.1).

Fig. 11.1 The STEM Play
Cycle

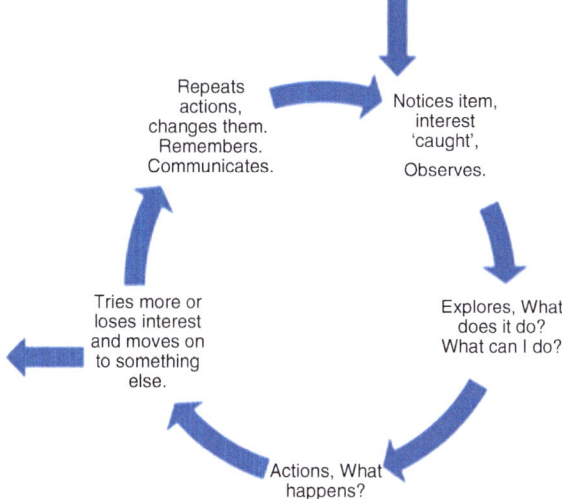

Progression in STEM Interactions

Watching the same children as they develop and recording actions at each stage
each time, the stages can reveal progress. Photographing the stages and pasting
them in a paper journal or online can also provide a record of instances and progress. Referring back with the children a year later can provide insights, especially
through using this photo journal.

Progress in interactions with the same objects or phenomena develop with age of
the child. The actions of three chdlren of increasing age as their actions progress
with interactions with water play are summarised in Table 11.1.

Identifying the stages in play has proved to be particularly useful to parents, and
several CASTME (ICASE member) groups are using it as the *CASTME Play Cycle*
and linking it with relevant photographs. This can be a useful tool for assessment of
the stages and are numbered in sequence (Table 11.2).

The Voice of the Child

As a biologist I observe, over time and most often visually. This elicits questions.
However, there are many types of observations that can be made to learn about the
child. The voice of the child is manifest in a number of ways. The play cycle
approach, whilst very useful and effective, is not the sole way of understanding a
child's development of interests, skills and understanding. Practitioners concerned
with the development and achievement of the very young beginners of STEM experiences often employ other techniques in striving to 'hear' the child's voice, and not

Table 11.1 The science (STEM) actions of three children (first encounter, a subsequent encounter by a 1-, 2- and 4-year-old): An observational study

Action sequence	Action	Experience	Science idea
Initial encounter Materials' basic properties	Child 1. 1 year old Hits water surface.	Exploring an unknown material.	Force, properties of material.
Exploring material	Child 2. 2 years Drops items into a water bowl: bath duck, metal spoon, wooden play block, pebble, bath sponge. ping-pong ball.	Experience of properties	Floating and sinking, absorption in the case of the sponge, which gives out water when squeezed. Force.
Changing – control over material	Child 3. 4 years. Collects things that he probably thinks would float or sink, and a small bottle with a lid.	Experimenting with items available, indicates some understanding of the properties of the material.	Pushing, floating and sinking.
Making something using previous knowledge	Child 3. Uses a lid (such as from toothpaste dispenser) to float as a boat, then fills it with water and it sinks.	Using previous experiential knowledge to fulfil child's planned objective.	Forces, knowledge of properties of boats. Experimenting with sinking hollow open objects.

Table 11.2 The stages of the STEM Play Cycle tabulated for progression monitoring

Play topic	Date/Time
	Name of child
Stage of STEM Play Cycle	Action of child in stage
Notices item – interest 'caught'.	
Observes.	
Explores. What does it do? What can I do?	
Actions. What happens?	
Tries more, or loses interest and does something else. Leaves this STEM Play Cycle.	
Repeats actions. Changes them.	
Remembers. Stores for future application.	
Communicates.	

the voices of adults directing them. We should recognise that combining different methods of listening to the child's 'voice' is encapsulated in the Mosaic approach (Clarke & Moss, 2001). The child's own voice can be 'listened' to by observing the body language when they encounter different phenomena, some of which may interest the child, others not. Hence, using such approaches, the science actions and interactions can be observed.

So how can we tell if they are interested and engaged? Tracking their path through any safe location can amplify this. In museum work, tracking visitors yields similar information on what catches their attention. Asking children to tell you about what they're doing and why may yield some interesting information; taking photographs of their activity, hearing their narrative as they play and talk aloud is informative and shows some interesting association in the child's mind between the actions and the object that they are manipulating. A 2-year-old's narrative reveals this. At one play session, a 2-year-old picked up the model female sheep, the ewe, and the model lamb. He put them side-by-side on the table. Then he rolled the ewe onto her back and said, *'Now the baby can get his mother's milk and not be hungry'.* His mother, when we told her, said that she was breastfeeding their new baby. When they have the opportunity, these very young children enjoy taking photographs on tablets and even mobile phones and will happily take interest in games and constructions. It is a particularly useful way of eliciting someone's interests, as has been found in museums, zoos and botanical gardens, and many parents and practitioners employ this technique in monitoring and keeping records of children's development. Such techniques reveal what is important to these children. Whilst there may be some commonality, it is important to bear in mind the cultures within which these children are experiencing their world and reacting to it. This may be very different from that to which we are used. Our way is not necessarily better, nor more appropriate, but it suits us.

Summary

In this chapter, I have discussed the meaning of science learning and traced its development to the present-day in English schools in particular. The need to recognise that pupils come to lessons with knowledge and experience which contribute to their engagement with subject matter of lessons has been highlighted. I acknowledge the importance of both formal and informal learning opportunities in a child's learning of science. It is particularly important for practitioners to be aware of the significance of learners' first encounter with science. This foundation awareness is in play and I discuss the importance of free choice play in developing scientific literacy. The sequence in such encounters with objects or phenomena is illustrated by the very simple STEM play cycle. Lastly, I stress the importance of practitioners recognising the child's voice and the relevance of understanding such in teaching.

References

Abrahams. I. (2021). Is science for all a flight of fancy?, *TES*, 9th April, 16–19. Retrieved from: www.tes.com

Carruthers, E., & Worthington, M. (2006). *Children's mathematics: Marking marks, making meaning*. Sage Publications.

Driver, R. (1983). *The pupil as scientist?* The Open University.

Eschach, H., & Fried, M. N. (2005). Should science be taught in early childhood? *Journal of Science Education and Technology, 14*(3), 315–336. https://doi.org/10.1007/s10956-005-7198-9

Gopnik, A. (2012). *The philosophical baby: What children's minds tell us about truth, life and the meaning of life*. Random House.

Milford, T., & Tippett, C. (2015). The design and validation of an early childhood STEM classroom observation protocol. *International Research in Early Childhood Education, 6*(1), 24–37. https://files.eric.ed.gov/fulltext/EJ1150965.pdf

Monteira, S. F., & Jiménez-Aleixandre, M. P. (2015). Practice of using evidence in kindergarten: The role of purposeful observation. *Journal of Research in Science Teaching, 53*(8), 1232–1258. https://doi.org/10.1002/tea.21259

Osborne, R. J., Bell, B. F., & Gilbert, J. K. (1983). Science teaching and children's views of the world. *European Journal of International Journal of Science Education, 23*(8), 847–862.

Riede, F., Johannsen, N., Hogberg, A., Nowell, A., Lombard, M., et al. (2018). The role of play objects and object play in human cognitive evolution and innovation. *Evolutionary Anthropology Issues News and Reviews, 27*, 46–59. https://doi.org/10.1002/evan.21555

Smith, W. S., McLaughlin, E., & Tunnicliffe, S. D. (1998). Effect on primary level students of Inservice teacher education in an informal science setting. *Journal of Science Teacher Education, 9*, 123–142. https://doi.org/10.1023/A:1009477616767

Tunnicliffe, S. D. (2020). The progression of children in learning about 'nature', our living world. In *Hands-on science. Science education. Discovering and understanding the wonders of nature* (pp. 216–220). Hands-on Science Network 2020. https://discovery.ucl.ac.uk/id/eprint/10105997/3/Tunnicliffe_Early%20years%20Science%20Biology%20HSCI%202020.pdf

Tunnicliffe, S. D., & Gkouskou, E. (2019). Science in action in spontaneous preschool play: An essential foundation for future understanding. *Early Child Development and Care, 91*(10), 54–63. https://doi.org/10.1080/03004430.2019.1653552

Vartiainen, J., & Kumpulainen, K. (2020). Playing with science: Manifestation of scientific play in early science inquiry. *European Early Childhood Education Research Journal, 28*(2), 1–14. https://doi.org/10.1080/1350293X.2020.1783924

Further Reading

Bruce, T., McNair, L., & Whinnett, J. I. (2021). *Putting the story telling at the heart of early childhood practice. A reflective guide for early years practitioners*. Routledge.

Clarke, A., & Moss, P. (2001). *Listening to young children: The mosaic approach*. National Children's Bureau for the Joseph Rowntree Foundation. http://www.jrf.org.uk/publications/listening-young

Dasgupta, P. (2021). *The economics of biodiversity: The Dasgupta review. Abridged version*. HM Treasury. www.gov.uk/official-documents

Driver, R., Guesne, E., & Tiberghien, A. (1985). *Children's ideas in science*. McGraw Hill.

Flewitt, R., & Ang, L. (2020). *Research methods for early childhood education*. Bloomsbury.

Gopnik, A. (2011). *The philosophical baby: What children's minds tell us about truth, life and the meaning of life*. Random House.

Journal of Emergent Science.. www.ase.org.uk

Katz, P. (2012). Using photobooks to encourage young children's science identities. *Journal of Emergent Science, 2*(3), 22–17.

Krapp, A. (1999). Interest, motivation, and learning: An educational-psychological perspective. *European Journal of Psychology of Education, 24*(1), 23–40. https://www.jstor.org/stable/23420114

Krapp, A., & Prenzel, M. (2011). Research on interest in science: Theories, methods, and findings. *International Journal of Science Education, 33*(1), 27–50.

McLeod, S. (2020). *Maslow's hierarchy of needs.*. https://www.simplypsychology.org/maslow.html

National Science Teachers Association (NSTA). https://www.nsta.org

Pollman, M. J. (2010). *Blocks and beyond. Strengthening Early Math and Science through Spatial Learning*. Paul. H. Brookes Publishing Co.

Pramling Samuelsson, I., & Asplund Carlsson, M. (2008). The playing learning child: Towards a pedagogy of early childhood. *Scandinavian Journal of Educational Research, 52*(6), 623–641.

Reiss, M. J., & White, J. (2014). An aims-based curriculum illustrated by the teaching of science in schools. *The Curriculum Journal, 25*, 76–89. https://doi.org/10.1080/09585176.2013.874953

Roberts-Holmes, G., & Moss, P. (2021). *Neoliberalism and early childhood education: Markets, imaginaries and governance*. Routledge.

Schiefele, U. (1991). Interest, Learning and motivation. *Educational Psychologist, 26*, 299–325. https://doi.org/10.1080/00461520.1991.9653136

Smith, P. K. (2010). *Children and play*. Wiley-Blackwell.

Tunnicliffe, S. D. (2016). *Starting inquiry based science in the early years. Look, talk, think and do. (Chapters 2 and 3)*. Routledge.

UNESCO. (2021). *Learn for our planet. A global review of how environmental issues are integrated in education*. Open Access. ISBN 978-92-3-100451-3.

Sue Dale Tunnicliffe, part of ICASE since 1988, is a zoologist specialising in education. Her doctorate (King's College London) was obtained after teaching across all ages; she set up and ran the Science and DT Advisory team for a London Authority and became Head of Education at the Zoological Society of London. She lectured in Winchester and Cambridge Universities before joining the Institute of Education, now UCK Faculty of Education and Society, where she is Reader in Science Education. She has published widely. She started the ICASE pre-secondary science section in 1988 and now is focusing on earliest years.

Chapter 12
Indigenous Knowledge and Science and Technology Education

Robby Zidny, Jesper Sjöström, and Ingo Eilks

Abstract In recent decades, research on the knowledge of indigenous cultures has gained more and more recognition in the field of science and technology education. Indigenous knowledge was promoted in terms of justice for indigenous peoples, respect for values and indigenous knowledge, the potential for intercultural learning, and chances for supporting education for sustainability. Various efforts have been made by both indigenous and non-indigenous scholars to introduce indigenous knowledge into science and technology curricula. In this chapter, we explore concepts and recent studies on indigenous knowledge in science and technology education, with important questions related to the topic.

Keywords Science education · Technology education · Indigenous knowledge · Ethnoscience · Teaching strategies · Education for sustainability · Systems thinking

R. Zidny
Department of Chemistry Education, Faculty of Teacher Training and Education, University of Sultan Ageng Tirtayasa, Serang, Indonesia
e-mail: robbyzidny@untirta.ac.id

J. Sjöström
Malmö University, Faculty of Education and Society, Department of Natural Science, Mathematics and Society, Malmö, Sweden
e-mail: jesper.sjostrom@mau.se

I. Eilks (✉)
University of Bremen, Department of Biology and Chemistry, Institute of Science Education, Bremen, Germany
e-mail: ingo.eilks@uni-bremen.de

© The Author(s), under exclusive license to Springer Nature Switzerland AG 2023
B. Akpan et al. (eds.), *Contemporary Issues in Science and Technology Education*, Contemporary Trends and Issues in Science Education 56, https://doi.org/10.1007/978-3-031-24259-5_12

What Is Indigenous Knowledge?

Indigenous knowledge (IK) can be understood as the knowledge, understanding, skills and philosophies held by local societies with long histories and experiences of interaction with their cultural and natural environments (Warren et al., 1993; Hiwasaki et al., 2014; Zidny et al., 2020). The definition of the term IK itself is initiating a debate among researchers. A definition that is generally agreed upon by all researchers is not available, and a variety of terminologies were used to characterize the knowledge of indigenous peoples (Berkes, 1993). Several different concepts and corresponding views are used to define IK. This distinction might be interpreted as a means by which to distinguish the ways of understanding nature among diverse cultural groups (Snively & Williams, 2016; Zidny et al., 2020). Some terms used to describe IK in the literature of science and technology education include indigenous science, ethnoscience, traditional ecological knowledge (TEK), native science, traditional (native) knowledge in general, or more specific, e.g., Yupiac science or Maori science, in particular. Among these terms, IK, indigenous science, TEK and ethnoscience were the most frequently used terms in the international literature. Recent reviews (e.g., Zidny et al., 2020) tried to clarify the terminologies and definitions related to IK to be used in science and technology education (Table 12.1).

IK has been recognized as an integral part of a complex system of knowledge that encompasses languages, sciences, systems of classification, resource use practice, cultural, philosophical and social interaction, and spirituality. This knowledge is crucial to cope with the world's cultural diversity and offers a basis for sustainable development in local regions (UNESCO, 2021). UNESCO has launched the Local and Indigenous Knowledge Systems (LINKS) program to promote and integrate IK in global climate science and policy. The LINKS program was intended to include IK in science-policy-society decisions with the role of how these knowledge systems may help us to understand, mitigate and adapt to climate change, prepare for natural disasters, prevent environmental degradation and biodiversity loss, and achieve sustainable development (see also Chap. 10 in this book). In terms of educational programs, UNESCO also suggests the transmission of IK into formal education. UNESCO programs promote the idea to bring IK into school curricula by initiating projects to move learning back into the community, such as the projects of the indigenous navigation book, introducing IK into science curricula in Vanuatu as a strategy for environmental conservation in local regions, and the development of curriculum and pedagogical materials that promote IK about the local flora and fauna of the Mayangna community in Nicaragua.

IK is a system of knowledge with unique and common features that share similarities to and differences from Western mainstream science (Stephens, 2000). De Beer et al. (2022) suggest an "intersecting perspective" within two knowledge systems which emphasize the shared tenets of IK and Western mainstream science. In terms of an organizing principle, IK adopts more holistic worldviews and integrates physical and spiritual perspectives, including moral aspects. IK also includes the inherited local wisdom as the habits of mind, which include the value to respect

Table 12.1 Terminologies and definitions related to IK in science education (Zidny et al., 2020)

No	Terminologies and acronyms	Definition	References
1	Indigenous (with an upper case I)	Refers to original inhabitants or first peoples in unique cultures who have experiences of European imperialism and colonialism.	Wilson (2008)
2	indigenous (with a lower case i)	Refers to things that have developed 'home-grown' in specific places.	Wilson (2008)
3	Indigenous knowledge	The local knowledge held by indigenous peoples or local knowledge unique to a particular culture or society.	Warren et al. (1993)
4	Indigenous science	The science-related knowledge of indigenous cultures. This science shaped indigenous knowledge based on the culture and perspective of an indigenous society.	Snively and Williams (2016); Kim and Dionne (2014)
5	Traditional ecological knowledge (TEK)	TEK is seen as part of IK, guided by indigenous scientific methods in parallel to Western modern science in terms of presenting solutions to ecological problems.	Kim and Dionne (2014)
6	Ethnoscience	Refers to a system of knowledge and cognition built to classify and interpret objects, activities, and events in a particular culture. Ethnoscience has been categorized into various disciplines of Western modern science-based scientific knowledge, namely ethnochemistry, ethnophysics, ethnobiology, ethnomedicine, and ethnoagriculture.	Sturtevant (1964); Abonyi et al. (2014)
7	Western science (WS)/Western Modern Science (WMS)/alternative Western thinking	Western Modern Science is understood as worldwide mainstream science, acknowledging that, also in modern Western societies, alternative worldviews and views on science and nature exist. This is called 'alternative Western thinking'.	Korver-Glenn et al. (2015)

nature, the living creatures and non-living objects. The knowledge transmission of IK is generally inherited through oral stories of the elders and through the cumulative experiences and observations of the community. The stories are connected to values, life experiences and proper behaviors. The knowledge system of IK is applied and integrated into daily living and traditional practices. Thus, learning about IK has potential to broaden learners' views on the Nature of Science (see Chap. 2 in this book).

Why Is IK Relevant to Science and Technology Education?

The integration of IK into science and technology education has become the concern of indigenous and non-indigenous scholars. Several major themes were frequently discussed in research and development concerning how IK is relevant to science and technology education (McKinley & Stewart, 2012; Zidny et al., 2020).

These include: (1) the justice of learning for students from indigenous backgrounds; (2) a potential contribution of IK to the base of mainstream Western modern science; (3) the integration of nature, philosophy and limits of science; (4) the potential of IK for intercultural learning, or (5) the role of IK in sustainability education.

The inclusion of IK into school curricula offers chances for more equity in science and technology education, which helps learners from indigenous backgrounds to understand the role of their cultural and societal context in the generation of scientific knowledge (Aikenhead & Michell, 2011). The recognition of IK in the classroom has the potential to reduce the assumption that learning science is 'strange' from the students' perspective (Mashoko, 2014). Equity for indigenous communities in education also recognizes the use of their frameworks and methodologies in terms of their history, policies, cultural and philosophical views (Smith, 1999).

The practice of IK in nature may open opportunities for scientists and citizens to consider IK as a potential solution facet for sustainability challenges, such as global diseases, climate change, environmental degradation and biodiversity loss (see also Chap. 10 in this book). Practices of indigenous communities are generally concerned with the sustainability of their nature, such as using only organic fertilizers and natural pesticides, crop rotation, conservation of water and soil, and anti-desertification practices (Atteh, 1989; Lalonde, 1993). The role of IK in promoting sustainability inspired educators to incorporate it into science and technology education in order to instill the value of nature conservation, critical self-reflection and education for sustainability (Parmin et al., 2017; Rahmawati et al., 2017; Zidny & Eilks, 2020).

IK has the potential to make a big contribution to the advancement of Western modern science. Scientists have started to explore IK in order to contribute to such science and technology fields as agriculture, medicine, chemistry and biology. For instance, researchers invested in discovering the potential of chemical natural product compounds and other sustainable materials that can benefit medicine or agriculture. Examples of these compounds include Azadirachtin as an insecticidal ingredient (Chaudhary et al., 2017), Aspirin as the medication to reduce inflammation (Rist & Dahdouh-Guebas, 2006) and Artemisinin as an anti-malarial agent (Klayman, 1985).

Research on the inclusion of cultural and philosophical aspects of IK in science and technology education has also been highlighted. The integration of IK in the classroom can expose various cultural backgrounds of the students, improve their interpretation of knowledge and make science learning more relevant (Botha, 2012; De Beer & Whitlock, 2009). Students can also obtain more experiences, foster their attitudes toward science and help preserve their values inherited from the local wisdom. The incorporation of IK into school curricula has the potential to enable students to gain further experiences, develop corresponding attitudes towards science, and promote intercultural learning (Kasanda et al., 2005; De Beer & Whitlock, 2009; Perin, 2011; Zidny et al., 2020; Zidny & Eilks, 2020).

Some important educational values can be gained by incorporating IK into science and technology education (Zidny & Eilks, 2022). The first is to encourage students to learn science across cultures, as urged by science organizations such as the International Union of Pure and Applied Chemistry (IUPAC) (Mahaffy, 2006). By providing the dual perspective on knowledge from IK and mainstream Western Modern Science, the students can open their insight into other cultures and philosophies in the application of science and technology (Zidny & Eilks, 2018, 2020; Zidny et al., 2021). Integration of IK in science curricula can also provide a rich and locally relevant context related to science, socio-cultural norms and philosophical values of society and, at the same time, offer alternative solutions to global challenges, such as conservation and adapting to climate change. Another benefit is to make the context of science learning more relevant for students in countries with, or near to, indigenous backgrounds (Zidny et al., 2020). The introduction of IK into science curricula can facilitate students to reflect the understanding of the different knowledge claims and worldviews (Botha, 2012; Fasasi, 2017). As a result, students in multicultural classrooms may find science learning more relevant and meaningful (Zidny et al., 2020; Zidny & Eilks, 2018). Another essential value is providing cross-disciplinary science learning to achieve education for sustainability and to foster systems thinking skills. Accordingly, as a result, learning science can help students to comprehend and appreciate how multiple disciplines, such as IK and mainstream WMS, should be involved in decision-making when it comes to addressing sustainability challenges (Koutalidi & Scoullos, 2016; Matlin et al., 2016; Zidny & Eilks, 2020).

How to Incorporate IK into Science and Technology Education?

There are some potential research areas on incorporating IK into science and technology education. One is concerned with empirical research on psychological and anthropological paradigms. These studies explore the cross-cultural experiences of students in the process of knowledge transition from the learner's environment into the science classroom (Aikenhead & Jegede, 1999). The research described the knowledge transition as 'cultural border crossing' (Aikenhead, 1996) and investigated cognitive problems that arise in various cultural environments (Jegede, 1995). The nature of the learner's initial ideas and beliefs on scientific phenomena was suggested in a cross-cultural context (Herbert, 2008).

Further research focuses on the development of instructional designs that introduce IK into science curricula, e.g., research aimed to help to solve cognitive conflicts among African students because of the differences between mainstream western modern science (WMS) in the school curricula and their cultural background (Abonyi, 1999). One study investigated the effect of ethnoscience-based instruction on students' concepts of scientific phenomena and attitudes towards science. Similar research using ethnoscience-based instruction was conducted by Fasasi (2017), involving the factor of school location and the education of the students by their parents. Instructional strategies were developed to construct an indigenous science education framework by involving the elders and the indigenous people in aboriginal communities (Aikenhead, 2001). Also, here, the students experienced a conflict when being faced with information from different knowledge systems (IK and WMS).

Other studies were concerned with the contextualization of IK in science and technology education. One contextualization has been suggested based on the framework of science in the form of traditional ecological knowledge (TEK) (Bermudez et al., 2017; Chandra, 2014; Hamlin, 2013). However, the concept of TEK tends to adopt a WMS view and has the risk of not sufficiently reflecting indigenous perspectives (Kim et al., 2017; Smith, 1999). It is suggested that one should thoroughly consider aspects such as socio-culture, philosophy, history and current policies of the indigenous community when incorporating indigenous knowledge in science and technology education (Kim et al., 2017).

A more recent focus is on research and development into the inclusion of IK in science and technology education by combining multiple perspectives of science learning from IK and WMS (Zidny et al., 2020, 2021; Zidny & Eilks, 2020). These studies produced a framework for how to elaborate on and design science education for sustainability that takes indigenous knowledge, WMS and alternative Western ideas into consideration. The studies suggest that science and technology education should reflect the different perspectives on sciences to achieve a more holistic worldview, intercultural understanding and sustainability.

So far, the inclusion of IK in science and technology education has been conducted by researchers in various regions, including Asia, Africa, Australia and the Americas. The Bridging the Gap (BTG) program was launched in Canada to offer students from Manitoba a culturally relevant curriculum and science-based environmental education. In Japan and Africa, Ogunniyi and Ogawa (2008) discussed the challenges of developing and implementing indigenous science curricula. The program of AKRSI (Alaska Rural Systemic Initiative) was conducted in Alaska to reconstruct IK and develop pedagogical instruction based on indigenous ways of the training of native Alaskan peoples in formal education (Barnhardt et al., 2000). The pedagogical approach combines learning processes inside and outside the classroom, which it is suggested helps to foster students' understanding and encourage them to learn about traditional values and culture (Barnhardt, 2007). In Australia, studies about IK were implemented in the higher education curricula. The findings imply that the IK-related curricula can help students to improve their critical reflection skills (Bullen & Roberts, 2019). In South-East Asia, especially in Indonesia, research on the reconstruction and integration of IK into science education and technology education started to become more recognized. Studies attempt to introduce science concepts behind IK, as well as the philosophy of local wisdom to emphasize the value of nature conservation and to develop critical reflection skills concerning the students' cultural backgrounds (Parmin et al., 2017; Rahmawati et al., 2017; Widiyatmoko et al., 2015). Later research in Indonesia aimed to explore the contexts and content of science learning from indigenous science of local communities (Zidny et al., 2021). The study suggests that IK can provide the student with relevant contexts to learn scientific concepts and offer an insight into its value in promoting sustainability. A case study on the implementation of a lesson plan integrating IK in schools and higher education courses was also conducted in Indonesia (Zidny & Eilks, 2020). The results indicated that the perception of the students towards the lesson was positive in terms of interest and relevance. The lesson showed the students that science lessons can be enriched with cross-disciplinary knowledge and worldviews to solve sustainability issues. In terms of the implementation of IK in the classroom, De Beer et al. (2022) suggest pre-and in-service teacher education to include the understanding of the tenets and knowledge both from IK and Western mainstream science.

In recent years, both indigenous and non-indigenous scholars have contributed to research and development into teaching methods, *didaktik models* (Sjöström et al., 2020), or strategies to integrate IK in science and technology education. Several teaching strategies on the inclusion of IK in the learning process were implemented, including ethnopedagogy (Burger, 1971), place-based teaching (Riggs, 2003), argumentation teaching (Ogunniyi, 2004), co-generative dialogue (Shady, 2014), tailored teaching (Hewson, 2000), culturally responsive teaching (Hernandez et al., 2013) or culturally responsive transformative teaching (CRTT) (Rahmawati et al., 2019). Other research on the inclusion of IK in science and technology education developed an educational research framework (Zidny et al., 2020) inspired by the Model of Educational Reconstruction (Duit et al., 2005) as one model for curriculum design in science and technology education (see Chap. 5 in this book). This framework suggests integrating multi-perspective views from IK and mainstream WMS into the educational reconstruction to foster sustainability education and intercultural learning. From this framework, two learning designs were derived by adapting sustainability-focused pedagogical approaches inspired by Burmeister et al. (2012). The first learning design adopted a sustainability-focused pedagogical approach, focusing on controversial sustainability issues in the case of the use of pesticides, and comparing the perspectives of IK and WMS (Zidny & Eilks, 2020). This learning design which introduces problem based learning is advocated to facilitate students to connect IK to the science in the classroom by exploring authentic and complex-structured problem (De Beer et al., 2022). The second learning design adopted a sustainability-focused approach, by using the context of indigenous chemistry (ethnochemistry), namely the use of natural bio-pesticides, to conduct investigations in green and sustainable chemistry (Zidny & Eilks, 2022).

For illustration, four teaching strategies should be described in brief, namely: tailored teaching (Hewson, 2000); culturally responsive teaching (Hernandez et al., 2013); culturally responsive transformative teaching (CRTT) (Rahmawati et al., 2019), and education for sustainability-based pedagogical approaches on the inclusion of IK (Zidny & Eilks, 2020, 2022).

Tailored Teaching

Hewson (2015) described tailored teaching strategies as a process of science teaching that should concern a suitable learning climate, including students' prior ideas into the lesson, using facilitation skills, fostering self-awareness, and adjusting the lesson to cope with students' needs. In tailored teaching, there are six steps of the learning process suggested: prepare, ask, give feedback, teach, apply and review. Hewson (2015) illustrated his tailored teaching model with a lesson plan concerning an animal (crocodile), integrating IK into a science course for primary schools, as follows:

- **Prepare:** A video or picture about crocodiles is shown. Visiting a museum with preserved crocodiles can be an option.
- **Ask:** Students discuss a story on interaction with and mythology about crocodiles.
- **Teach:** Local indigenous people familiar with crocodiles are invited to talk about them. Students research information on crocodiles, and record their research by drawing or photographing aspects of crocodiles in their habitats.
- **Apply:** A project of developing sustainability in crocodile farming is assigned to the students.
- **Review:** Students create a picture of the crocodile with correct labels, then make reports about their findings on the management and ecology of crocodile farming.

Culturally Responsive Teaching (CRT)

Hernandez et al. (2013) suggested the culturally responsive teaching (CRT) model using a qualitative theoretical study from the literature of CRT practices. The model of CRT encompasses five main components that should be included in science teaching and learning: (i) content integration; (ii) facilitating knowledge construction; (iii) prejudice reduction; (iv) social justice; and (v) academic development:

- **Content integration:** Integrating information from different cultures (for example, from IK and WMS) and making connections to the students' daily life experiences.
- **Facilitating knowledge construction:** Helping the students to reconstruct their prior knowledge as a means to make science learning accessible.
- **Prejudice reduction:** Using local or indigenous language to promote positive interaction between students.
- **Social justice:** Encouraging students to challenge the *status quo* to develop socio-political or critical thinking.
- **Academic development:** Using various methods or modelling to create learning opportunities.

Culturally Responsive Transformative Teaching (CRTT)

Culturally Responsive Transformative Teaching (CRTT) is the result of the further development of the CRT model (Rahmawati et al., 2019). In this model, five steps of the teaching and learning process are involved, which include: (i) self-identification; (ii) cultural understanding; (iii) collaboration; (iv) critical reflective thinking; and (v) transformative construction:

- **Self-identification:** Students conduct self-reflection to develop their identities within personal differences.
- **Cultural understanding:** Students participate in cultural understanding and knowledge construction through various resources (articles, videos, etc.) about IK.
- **Collaboration:** Students discuss concepts and cultural perspectives in groups.
- **Critical reflective thinking:** Students participate in debates to explore different perspectives and reflect on their understanding and values.
- **Transformative construction:** Students transform their values and understandings and present their learning in new ways.

Sustainability Education-Based Pedagogical Approaches

Two learning designs on the inclusion of IK in science and technology education were developed based on sustainability education-based pedagogical approaches (Zidny & Eilks, 2020, 2022). The first learning design was used in a lesson plan that focuses on the discussion of the controversial issue of pesticides use:

- Exploring prior personal knowledge: Brainstorming student ideas about the topic of pesticides' use and identifying their initial thinking and opinion.
- Introducing the context and the controversial issue: Exploring the context of pesticides' use from various resources and provoking controversial questions about such use.
- Initiating multi-perspective thinking from IK and WMS: Identifying the scientific phenomena, philosophy, values and local wisdom from IK to complement WMS views as well as finding alternative solutions from both perspectives.
- Connecting the context with chemistry concepts: Making the connection between the issue and the relevant concepts of science.
- Meta-reflection: Students consider consequences and the potential collaboration, as well as reflecting the process of decision-making and future deliberations toward the issue.

The second learning design was implemented in a lesson plan to facilitate the students to learn green and sustainable chemistry using the example of pesticides' use inspired by IK:

- Exploring ethnochemistry and related ideas from green chemistry: Exploring information from IK and the concept of green chemistry related to pesticides' use from various resources.
- Understanding the chemistry concepts behind ethnochemistry: Investigating the chemical concept behind the use of natural products inspired by IK and discussing the findings.
- Deepening the science concepts behind ethnochemistry through various chemical experiments: Using experiments to investigate the chemical compounds that can be used as alternative green and sustainable pesticides.
- Discussion of experimental results: Students discuss the results of the experiments in a group discussion.
- Evaluating laboratory methods based on green chemistry: Students evaluate and compare the optimal methods that mostly represent green chemistry.

What Do We Know About the Effects?

There is growing interest in and involvement of global educators, researchers and organizations in the inclusion of IK in science and technology education. The challenge now is to create a conceptual bridge between IK and mainstream WMS to make synergy for the welfare of society and strengthen science and technology education (Hewson, 2015). The collaboration between IK and mainstream WMS can be used as the basis for decision-making in education, natural resource management and sustainability. The important contribution of IK to global development has been recognized as a problem-solving strategy to handle sustainability issues such as climate change, biodiversity loss and environmental degradation. The concern that should be taken into account by educators is to learn and find a way to deliver and integrate IK in the classrooms.

Frameworks, methods, models or strategies of teaching for the inclusion of IK are diverse and challenging to apply in the classroom. Their application is not only relevant for students with indigenous backgrounds, but also for students with non-indigenous backgrounds in heterogeneous classrooms. The choice of ways to deliver the learning context or content should take into consideration the student interests and needs and reflect on which knowledge, skills, and attitudes should be infused in the processes of teaching and learning.

In terms of curriculum development, there is a strong chance of IK being integrated. IK can be integrated into the different visions of scientific literacy in science and technology education (Sjöström & Eilks, 2018). Vision 1 concerns the learning of scientific content and processes for future application; Vision 2 focuses on understanding the utilization of scientific knowledge in everyday life and social contexts, while Vision 3 emphasizes philosophical values, transdisciplinarity and critical thinking. In this case, IK perspectives can extend Vision 3 by providing transdisciplinary aspects and multiple perspectives on scientific worldviews, which can further complement mainstream WMS (Zidny et al., 2020; Murray, 2015; Sjöström, 2018).

Reflecting on the research on the inclusion of IK into science and technology education, educators and researchers need to explore more about the context or content of IK that can be used and reconstructed to fit with the curriculum and students' needs. This effort should be carried out carefully and should respect the culture of indigenous peoples. In the reconstruction process, research should consider any aspects of socio-cultural, philosophical, political and local wisdom of indigenous communities. The integration of IK into the formal science classroom is indeed a major challenge, since we face a growing diversity in cultural backgrounds of the students in many countries. Teaching methods, models or strategies that were successfully implemented in a region or country might be not compatible in another country with different characteristics of culture or society. More research and development is needed to find appropriate and effective teaching and learning approaches to integrate IK in science and technology education.

Summary

In this chapter the nature of indigenous knowledge (IK) and its relevance to science and technology education are discussed. It is outlined which suggestions are made to incorporate IK into science and technology education and what are potential effects on science and technology teaching and learning.

Recommended Resources

- The UNESCO *Local and Indigenous Knowledge Systems (LINKS) Project.* https://en.unesco.org/links
- Indigenous Knowledge in Global Policies and Practice for Education, Science and Culture by Douglas Nakashima (2010) https://unesdoc.unesco.org/ark:/48223/pf0000265855
- Living knowledge. Indigenous knowledge in science education https://living-knowledge.anu.edu.au/index.htm
- Considering the Value of Indigenous Knowledge and Practices by Sara Krauskopf (2021) https://www.nsta.org/science-teacher/science-teacher-septemberoctober-2021/considering-value-indigenous-knowledge-and
- Discussion on Decolonizing Science Education & Practicing Indigenous Science: Dialogue 13 (2021) https://www.waysofknowingforum.ca/dialogue-13
- The Conversation: Why indigenous knowledge has a place in the school science curriculum (2015) https://theconversation.com/why-indigenous-knowledge-has-a-place-in-the-school-science-curriculum-44378

References

Abonyi, O. S. (1999). *Effects of an ethnoscience-based instructional package on students' conception of scientific phenomena and interest in science* (Thesis). University of Nigeria Nsukka.

Abonyi, O. S., Njoku, L. A., & Adibe, M. I. (2014). Innovations in science and technology education: A case for ethnoscience based science classrooms. *International Journal of Scientific & Engineering Research, 5*(1), 52–56.

Aikenhead, G. S. (1996). Science education: Border crossing into the subculture of science. *Studies in Science Education, 27*, 1–52.

Aikenhead, G. S. (2001). Integrating Western and aboriginal sciences: Cross-cultural science teaching. *Research in Science Education, 31*, 337–355.

Aikenhead, G. S., & Jegede, O. J. (1999). Cross-cultural science education: A cognitive explanation of a cultural phenomenon. *Journal of Research in Science Teaching, 36*, 269–287.

Aikenhead, G. S., & Michell, H. (2011). *Bridging cultures: Indigenous and scientific ways of knowing nature.* Pearson.

Atteh, O. D. (1989). *Indigenous local knowledge as key to local level development: Possibilities, constraints, and planning issues in the context of Africa.* Unpublished manuscript.

Barnhardt, R. (2007). Creating a place for indigenous knowledge in education: The Alaska native knowledge network. In G. Smith & D. Gruenewald (Eds.), *Place-based education in the global age: Local diversity* (pp. 113–133). Lawrence Erlbaum.

Barnhardt, R., Kawagley, A. O., & Hill, F. (2000). Educational renewal in rural Alaska. In *Proceedings of the International Conference on Rural Communities & Identities in the Global Millennium* (pp. 140–145).

Berkes, F. (1993). Traditional ecological knowledge in perspective. In J. T. Inglis (Ed.), *Traditional ecological knowledge: Concepts and cases* (pp. 1–9). International Development Research Centre (IRDC) Books.

Bermudez, G. M. A., Battistón, L. V., García Capocasa, M. C., & de Longhi, A. L. (2017). Sociocultural variables that impact high school students' perceptions of native fauna: A study on the species component of the biodiversity concept. *Research in Science Education, 47*, 203–235.

Botha, L. R. (2012). Using expansive learning to include indigenous knowledge. *International Journal of Inclusive Education, 16*(1), 57–70.

Bullen, J., & Roberts, L. (2019). Driving transformative learning within Australian indigenous studies. *The Australian Journal of Indigenous Education, 48*, 12–23.

Burger, H. G. (1971). *Ethno-pedagogy: A manual in cultural sensitivity with techniques for improving cross-cultural teaching by fitting ethnic patterns* (2nd ed.). Southwestern Cooperative Educational Lab.

Burmeister, M., Rauch, F., & Eilks, I. (2012). Education for sustainable development (ESD) and chemistry education. *Chemistry Education Research and Practice, 13*, 59–68.

Chandra, D. V. (2014). Re-examining the importance of indigenous perspectives in the Western environmental education for sustainability: From tribal to mainstream education. *Journal of Teacher Education for Sustainability, 16*, 117–127.

Chaudhary, S., Kanwar, R. K., Sehgal, A., Cahill, D. M., Barrow, C. J., Sehgal, R., & Kanwar, J. R. (2017). Progress on Azadirachta indica-based biopesticides in replacing synthetic toxic pesticides. *Frontiers in Plant Science, 8*(610), 1–13.

De Beer, J., & Whitlock, E. (2009). Indigenous knowledge in the life sciences classroom: Put on your de bono hats! *The American Biology Teacher, 71*, 209–216.

De Beer, J., Petersen, N., & Ogunniyi, M. (2022). Indigenous knowledge in science education: Implications for teacher education. In J. A. Luft & M. G. Jones (Eds.), *Handbook of research on science teacher education* (pp. 340–351). Routledge.

Duit, R., Gropengießer, H., & Kattmann, U. (2005). Towards science education that is relevant for improving practice: The model of educational reconstruction. In H. Fischer (Ed.), *Developing standards in research on science education* (pp. 1–9). Taylor & Francis.

Fasasi, R. A. (2017). Effects of ethnoscience instruction, school location, and parental educational status on learners' attitude towards science. *International Journal of Science Education, 39,* 548–564.

Hamlin, M. L. (2013). "Yo soy indígena": Identifying and using traditional ecological knowledge (TEK) to make the teaching of science culturally responsive for Maya girls. *Cultural Studies of Science Education, 8,* 759–776.

Herbert, S. (2008). Collateral learning in science: Students' responses to a cross-cultural unit of work. *International Journal of Science Education, 30,* 979–994.

Hernandez, C. M., Morales, A. R., & Shroyer, M. G. (2013). The development of a model of culturally responsive science and mathematics teaching. *Cultural Studies of Science Education, 8,* 803–820.

Hewson, M. G. (2000). A theory-based faculty development program for clinician-educators. *Academic Medicine, 75,* 498–501.

Hewson, M. G. (2015). Integrating indigenous knowledge with science teaching. In M. G. Hewson (Ed.), *Embracing indigenous knowledge in science and medical teaching* (pp. 119–131). Springer.

Hiwasaki, L., Emmanuel, L., Syamsidik, S., & Shaw, R. (2014). *Local and indigenous knowledge for community resilience: Hydro-meteorological disaster risk reduction and climate change adaptation in coastal and small island communities.* UNESCO. Retrieved from: https://unesdoc.unesco.org/ark:/48223/pf0000228711

Jegede, O. (1995). Collateral learning and the eco-cultural paradigm in science and mathematics education in Africa. *Studies in Science Education, 25,* 97–137.

Kasanda, C., Lubben, F., Gaoseb, N., Kandjeo-Marenga, U., Hileni Kapenda, H., & Campbell, B. (2005). The role of everyday contexts in learner-centred teaching: The practice in Namibian secondary schools. *International Journal of Science Education, 27,* 1805–1823.

Kim, E. J. A., & Dionne, L. (2014). Traditional ecological knowledge in science education and its integration in grades 7 and 8 Canadian science curriculum documents. *Canadian Journal of Science, Mathematics and Technology Education, 14,* 311–329.

Kim, E.-J. A., Asghar, A., & Jordan, S. (2017). A critical review of traditional ecological knowledge (TEK) in science education. *Canadian Journal of Science, Mathematics and Technology Education, 17,* 258–270.

Klayman, D. L. (1985). Qinghaosu (artemisinin): An antimalarial drug from China. *Science, 228*(4703), 1049–1055.

Korver-Glenn, E., Chan, E., & Howard Ecklund, E. (2015). Perceptions of science education among African American and white evangelicals: A Texas case study. *Review of Religious Research, 57,* 131–148.

Koutalidi, S., & Scoullos, M. (2016). Biogeochemical cycles for combining chemical knowledge and ESD issues in Greek secondary schools part I: Designing the didactic materials. *Chemistry Education Research and Practice, 17,* 10–23.

Krauskopf, S. (2021). Considering the value of indigenous knowledge and practices. *The Science Teacher, 89*(1), 8–10.

Lalonde, A. (1993). African indigenous knowledge and its relevance to sustainable development. In J. Inglis (Ed.), *Traditional ecological knowledge: Concepts and cases* (pp. 55–62). Canadian Museum of Nature.

Mahaffy, P. (2006). Moving chemistry education into 3D: A tetrahedral metaphor for understanding chemistry: Union carbide award for chemical education. *Journal of Chemical Education, 83,* 49–55.

Mashoko, D. (2014). Indigenous knowledge for plant medicine: Inclusion into school science teaching and learning in Zimbabwe. *International Journal of English and Education, 33,* 2278–4012.

Matlin, S. A., Mehta, G., Hopf, H., & Krief, A. (2016). One-world chemistry and systems thinking. *Nature Chemistry, 8*(5), 393–398.

McKinley, E., & Stewart, G. M. (2012). Out of place: Indigenous knowledge (IK) in the science curriculum. In B. Fraser, K. Tobin, & C. McRobbie (Eds.), *Second international handbook of science education* (pp. 541–554). Springer.

Murray, J. J. (2015). Re-visioning science education in Canada: A new polar identity and purpose. *Education Canada, 55*(4), 18–21. Retrieved from: http://www.cea-ace.ca/education-canada/article/re-visioning-science-education-Canada

Nakashima, D. (2010). *Indigenous knowledge in global policies and practice for education, science and culture. Chief, Small Islands and Indigenous Knowledge Section, Natural Sciences Sector*. UNESCO.

Ogunniyi, M. B. (2004). The challenge of preparing and equipping science teachers in higher education to integrate scientific and indigenous knowledge systems for their learners. *South African Journal of Higher Education, 18*, 289–304.

Ogunniyi, M. B., & Ogawa, M. (2008). The prospects and challenges of training South African and Japanese educators to enact an indigenized science curriculum. *South African Journal of Higher Education, 22*, 175–590.

Parmin, S., Ashadi, S., & Fibriana, F. (2017). Science integrated learning model to enhance the scientific work independence of student teachers in indigenous knowledge transformation. *Journal Pendidikan IPA Indonesia, 6*, 365–372.

Perin, D. (2011). Facilitating student learning through contextualization: A review of evidence. *Community College Review, 39*, 268–295.

Rahmawati, Y., Ridwan, A., & Nurbaity. (2017). *Should we learn culture in chemistry classroom? Integration ethnochemistry in culturally responsive teaching*. AIP Conference Proceedings, 1868.

Rahmawati, Y., Ridwan, A., Rahman, A., & Kurniadewi, F. (2019). Chemistry students' identity empowerment through ethnochemistry in culturally responsive transformative teaching (CRTT). *Journal of Physics: Conference Series, 1156*(1), 012032.

Riggs, E. M. (2003). Field-based education and indigenous knowledge: Essential components for geoscience teaching for Native American communities. *Science Education, 89*, 296–313.

Rist, S., & Dahdouh-Guebas, F. (2006). Ethnosciences – A step towards the integration of scientific and indigenous forms of knowledge in the management of natural resources for the future. *Environment, Development and Sustainability, 8*, 467–493.

Shady, A. (2014). Negotiating cultural differences in urban science education: An overview of teachers' first-hand experiences: Reflection of cogen journey. *Cultural Studies of Science Education, 9*, 35–51.

Sjöström, J. (2018). Science teacher identity and eco-transformation of science education: Comparing Western modernism with Confucianism and reflexive Bildung. *Cultural Studies of Science Education, 13*, 147–161.

Sjöström, J., & Eilks, I. (2018). Reconsidering different visions of scientific literacy and science education based on the concept of Bildung. In Y. Dori, Z. Mevarech, & D. Baker (Eds.), *Cognition, metacognition, and culture in STEM education* (pp. 65–88). Springer.

Sjöström, J., Eilks, I., & Talanquer, V. (2020). Didaktik models in chemistry education. *Journal of Chemical Education, 97*, 910–915.

Smith, L. T. (1999). *Decolonizing methodologies: Research and indigenous peoples*. Zed Books.

Snively, G., & Williams, W. L. (2016). *Knowing home: Braiding indigenous science with Western science, book 1*. University of Victoria.

Stephens, S. (2000). *Handbook for culturally responsive science curriculum*. Alaska Native Knowledge Network.

Sturtevant, W. C. (1964). Studies in ethnoscience. *American Anthropologist, 66*(3), 99–131.

UNESCO. (2021). *Local and Indigenous Knowledge Systems (LINKS)*. https://en.unesco.org/links

Warren, D. M., Brokensha, D., & Slikkerveer, L. J. (Eds.). (1993). *Indigenous knowledge systems: The cultural dimension of development*. Kegan Paul.

Widiyatmoko, A., Sudarmin, S., & Khusniati, M. (2015). Reconstruct ethnoscience based-science in karimunjawa islands as a mode to build nature care student character. In *International Conference on Mathematics, Science, and Education 2015 (ICMSE 2015)* (pp. SE 65–SE 70).

Wilson, S. (2008). *Research is ceremony: Indigenous research methods.* Fernwood.

Zidny, R., & Eilks, I. (2018). Indigenous knowledge as a socio-cultural context of science to promote transformative education for sustainable development: A case study on the Baduy community (Indonesia). In I. Eilks, S. Markic, & B. Ralle (Eds.), *Building bridges across disciplines for transformative education and a sustainable future* (pp. 249–256). Shaker.

Zidny, R., & Eilks, I. (2020). Integrating perspectives from indigenous knowledge and Western science in secondary and higher chemistry learning to contribute to sustainability education. *Sustainable Chemistry and Pharmacy, 16*, 100229.

Zidny, R., & Eilks, I. (2022). Learning about pesticides use adapted from ethnoscience as a contribution to green and sustainable chemistry education. *Education Sciences, 12*(4), 227.

Zidny, R., Sjöström, J., & Eilks, I. (2020). A multi-perspective reflection on how indigenous knowledge and related ideas can improve science education for sustainability. *Science & Education, 29*(1), 145–185.

Zidny, R., Solfarina, S., Aisyah, R. S. S., & Eilks, I. (2021). Exploring indigenous science to identify contents and contexts for science learning in order to promote education for sustainable development. *Education Sciences, 11*(3), 114.

Dr. Robby Zidny is a lecturer and researcher at the Department of Chemistry Education in the Faculty of Teacher Training, University of Sultan Ageng Tirtayasa, Indonesia. In 2021, he received his PhD in chemistry education from the University of Bremen, Germany. He has 8 years of teaching experience in chemistry education courses, such as general chemistry, biochemistry, environmental chemistry, or methodology of chemistry education research. He also has experience in teacher continuous professional development for in-service chemistry teachers. The focus of his research is integrating indigenous science in chemistry education to promote education for sustainability and green and sustainable chemistry education. In 2020, he received a research grant from the German Academic Exchange Service (DAAD) to conduct research on integrating indigenous science in chemistry education to promote green and sustainable chemistry.

Prof. Dr. Jesper Sjöström has been a full professor of science education at Malmö University, Sweden since 2021. He was employed as a senior lecturer (Assistant Professor) at Malmö University in 2007 and was promoted to Associate Professor in 2015. He is a fully trained upper-secondary school teacher oriented towards chemistry and general science/science studies. He also has 3 years of experience from the research area of Science and Technology Studies (STS) as a post-doc researcher at Lund University, Sweden. Two important concepts in his research are eco-reflexive *Bildung* and *Didaktik*. Related to this, he is interested in *Bildung*-oriented *didaktik* models and modelling, as well as in the philosophy of science education. An important framing for him is the Anthropocene era and related global environmental and socio-political issues. Based on this, he is asking what an eco-reflexive *Bildung*-oriented science (and especially chemistry) education might look like?

Prof. Dr. Ingo Eilks, FRSC has been a full professor in chemistry education at the Institute of Science Education at the University of Bremen, Germany since 2004. Having been a grammar school teacher for chemistry and mathematics, he engaged in the domain of chemistry education research and development. His research interests encompass socio-scientific issues-based science teaching, education for sustainability, learning with digital media, teaching the particulate nature of matter, or co-operative learning, all for high school and tertiary levels. For more than 20 years, these interests have been deeply connected to co-operatively working with teachers following participatory action research in science education. Further research focuses teachers' beliefs and learning as well as innovations in higher chemistry education. He has received different national and international awards for research and teaching, among them the 2017 Award for Outstanding Contributions to the Incorporation of Sustainability into Chemical Education of the American Chemical Society – Committee for Environmental Improvement.

Chapter 13
Public Understanding of Science and Technology

Janchai Yingprayoon

Abstract In this article, the concept of public understanding of science and technology in general is discussed, including how to proceed, as well as government policies to present the popularization of science to the public so that people understand and know how to apply science to help solve everyday life problems. The article also presents the concepts of scientific tourism, how to organize science activities in schools, science projects and science camps.

Keywords Public understanding · Science popularization · Public lecture · Science talk show · Project 2000+ · Science museum · Science center · Scientific tourism · Science project · Science club · Science camp

Concept of Public Understanding of Science and Technology

Our world is constantly changing in a dynamic way, in terms of life as a whole: living, eating, traveling or even in terms of culture, etc. All changes are the result of scientific and technological advances. In this context, science covers mathematics, technology, engineering and medicine.

Will our world be better or different? The key factor is knowledge and understanding of science. This should be an important element in promoting self-improvement of the people and the development of every country on this planet.

In time of world crisis, all people must be ready for preparedness and learning through science in order to face difficulties and adapt themselves to change. As all of us can witness, the COVID-19 pandemic is an evident example of crises causing major changes in people's way of life and has widely affected the economy, society or even culture of communities around the world.

J. Yingprayoon (✉)
Demonstration School of Thaksin University, Thaksin University, Phatthalung Campus, Phatthalung, Thailand

© The Author(s), under exclusive license to Springer Nature Switzerland AG 2023
B. Akpan et al. (eds.), *Contemporary Issues in Science and Technology Education*, Contemporary Trends and Issues in Science Education 56, https://doi.org/10.1007/978-3-031-24259-5_13

From adaptation to survival, people need to know and understand the science that directly affects their personal being: disease prevention, vaccination, public health, food, contact with others, travel, or even online trading services, transportation of goods, etc. The COVID-19 pandemic represents a crisis that has caused some of the most significant changes in the history of mankind and has had a strong impact on human life, in various aspects. Understanding science is a way to enable people to efficiently protect themselves and other members of the society from the pandemic.

In addition to the COVID-19 epidemic, we are also dealing with energy-related problems. As we know, many countries are phasing out the use of gasoline-powered cars and are turning to electric-powered vehicles. People should acquire a certain level of knowledge and understanding of science to be familiar with this issue and to face the problematic situations that will follow.

Changes certainly require scientific knowledge, in many dimensions, to help to solve the problems caused. With regard to commercial and industrial business sectors, investors, executives and operators also need to have knowledge and understanding of science to achieve real success and overcome their competitors.

Ignorance of the importance of knowledge and understanding of science could cause a nation to lag behind and face economic depression and livelihood difficulties later on.

Governments must play a major role in putting in place policy guidelines enabling the people to understand science. Governments should also make judgements that are based on scientific understanding regarding those projects requiring community participation that are agreed upon and presented by the people.

As a matter of fact, more and more issues related directly to people's lives should be tackled with science, including environmental problems, pollution, radioactive waste disposal, adjustment of fluoride to tap water, vaccination against various diseases, appropriate use of land and fishing grounds.

In some cases, collective decision-making is needed, as these issues involve many parties and become public issues having an effect on the lives of human beings and animals, as well as on property in the long run. Such public issues include pregnancy problems, abortion, nuclear hazards, acid rain caused by industrial plants, smoking in public spaces, or science experiments on animals.

In addition, governments must foster a public understanding of science and technology. They should prioritize an educational policy to equip students with fundamental scientific knowledge enabling them to develop critical and analytical thinking. This educational policy should also encourage students into lifelong learning.

In 1993, ICASE and UNESCO jointly organized the World Forum Project 2000+: *Scientific and Technological Literacy for all* (STL), aimed at equipping the global population with scientific knowledge to improve life and build a new and better society. Following this world forum, different projects were implemented in many countries to develop scientific popularity.

Science Popularization

Science popularization is an effective tool and strategy for communicating science to people (Danilina, 2022). Disseminating scientific knowledge means bringing science to the general public. The dissemination of scientific knowledge is a powerful and strategic measure for shaping modern society.

The dissemination of knowledge not only involves the knowledge and skills that are useful to people's daily lives; it also covers general guidelines and the culture of the society concerned. In general, the conflicts arising between scientific communities and public opinions are linked to distrust and doubt among the population. There is also the bias shown by scientists or workers towards the general public to consider. Many scientists believe that the public lack adequate scientific knowledge to understand the existing problems. Moreover, obstacles often arise from misunderstanding due to the use of scientific and technological terminologies, or the communication of erroneous facts. Consequently, competent people, with a solid knowledge of science, are needed to provide the public with an accurate understanding of scientific truth behind the problems concerned. In this respect, the above-mentioned obstacles could be overcome and the mission successfully achieved.

In fact, conflicts or misunderstandings can arise in the conducting of any project. However, it is important to avoid such obstacles as much as possible, so as to ensure the success of science knowledge dissemination to the general public.

If we could make science popular among the general public, change for the betterment of society through the use of science would not be difficult task. Society has evolved in a way that quickly enhances a better standard of living. Decision-making when dealing with society's problems as a whole could be made easier as well.

Nowadays, scientific popularity can be created through social media, which people of all ages can easily and effectively access. Apart from acknowledging information from those in charge of the dissemination of scientific knowledge, exchange of information and opinions among people around the globe is proving to be effortless when using social media.

Nevertheless, we should be aware of the possibility of fake news, or false data, provided through social media. Distorted scientific information definitely affects the popularity and accurate knowledge of science. Thus, people should understand the basic science to a certain extent so as to be able to reflect, analyze any issues and make an appropriate and reasonable judgement when problem-solving.

The promotion of scientific popularity among the world's population can be carried out in different ways, depending on variables such as age, gender and taste, as well as cultural beliefs that have been cultivated from the past.

Popularization of science may be achieved through educational and entertaining events, such as public science lectures or science talk shows, and science entertainment festivals for children and families. In this connection, science museums or science centers, contemporary libraries with modern technology, playful atmosphere and easy access also play an important role.

Public Lecture and Science Talk Show

In a learning society, public lectures on scientific developments are often organized to promote scientific popularity and educate both adults and students of various ages. These lectures can be conducted in various forms, using materials and communication techniques that make them easily accessible to the public and to attract the attention of the visitors, for example, TED talks (Sugimoto et al., 2013).

Lectures or science performances may take place at times corresponding to important national events, such as Christmas Day, Teachers' Day, National Science Week, Labor Day, etc. Special activities can also be organized on the occasion of annual meetings of scientific professional associations.

In many countries it is considered traditional to hold lectures on specific days to commemorate the national scientists of that nation. Science shows can be provided by the private sector with a view to educate and entertain through science. Science shows can be performed to provide knowledge about light and shadow, weather, environmental conservation, or even about science toys.

Government sectors may contribute to science lectures and performances by funding private scientific companies or educational institutions to encourage community events across the country. Lectures or demonstrations of science can be held via social media, which offer many choices of channels and platforms.

During the COVID-19 outbreak, gatherings have not always been possible. Communication and learning online have thus proved to be necessary. We can see that many more science lectures and demonstrations have been organized online than ever before. In fact, dissemination of knowledge through online learning does not focus solely on scientific subjects, but covers all the other subjects programmed in the school curriculum, as students have had to learn from home.

The COVID-19 crisis has made it clear that the new method of teaching and learning has changed the whole world. Dissemination of knowledge does not concentrate only on the subjects normally taught in school, but also includes the use of information technology in daily life to avoid the spread of the COVID-19 virus within schools. Some schools implement a policy of using QR codes or a barcode scan on smartphones, instead of using cash. Due to their popularity, smartphones have become an essential tool in today's world. Information technology plays a vital role in shopping, banking business and travel. Those people without knowledge of or access to information technology systems will certainly face difficulties with living in modern society.

Science Museums or Science Centers

In the past, science museums or science centers were suited to the role of disseminating scientific knowledge (Georgopoulou et al., 2021; Gunay, 2012). They usually featured permanent exhibitions about the history of nature, archaeology, geology, industry and industrial tools, among others.

Modern science museums tend to be distinctively different from the old versions. Activities in modern science museums can be held in three forms: permanent, temporary, or mobile science exhibitions (MSE). Many science museums also have a planetarium to educate the public about the Earth and stars, or to introduce eccentric objects in a way to arouse public interest. Presentation of information is conducted in a wide context, together with interactive forms of activity, generally in three dimensions. This means that the mission of traditional science museums has shifted towards a more technological approach and dissemination of knowledge. Several science museums have been renamed 'science centers', while some have become 'discovery centers'.

Many science museums consider the family relationship to be important and usually organize family activities to provide knowledge about science in daily life. Science birthday parties can also be arranged. Some science museums in Europe that possess an interesting history organize exhibitions about that history as a way to disseminate knowledge of science and nature at the same time.

A new concept of science museum consists of a combination of science with culture. Modern science museums often combine basic school education with the advancement of modern science. Some science museums provide a science job offer corner and consulting services for visitors who are interested in working in scientific sectors.

At present, a network of international science museums has been created to enable the exchange of experiences in communication and dissemination of science knowledge and technology to the public. Conferences, seminars, workshops and competitions on science projects for students are also organized at national and international levels, within the framework of this network.

Scientific Tourism

Another concept in disseminating scientific knowledge to the public focuses on scientific tourism. Various beautiful sites can be used as sources for learning about nature and science, in enjoyable and pleasant surroundings. Scientific tourism can be conducted in many forms to enable the learning about the human relationship with animals and plants. This usually leads to an awareness about the preservation of flora and fauna.

Institutional Tourism

This is classified as tourism for those of the general public who wish to acquire knowledge about science. Various institutions offer these kinds of visits, including science museums and science centers.

Eco Tourism

This type of tourism emphasizes the importance of ecology and ecological systems, as well as the awareness of and the public responsibility for the ecological impact of society.

Natural Tourism

This kind of excursion consists of a trip to learn about natural sites such as seas, islands, mountains, caves, exploring flora and fauna in different environments, land and aquatic animals, freshwater and marine animals.

Natural Phenomena Tourism

Traveling to observe natural phenomena makes for an interesting excursion. This may consist of the observation of solar movement or phenomena in the sky on special occasions. Interestingly, legends related to the sky are often linked to local lifestyles or local cultures. This fact is worth emphasizing to encourage further study and research to understand the evolution of local societies.

Health Tourism

At present, the importance of health tourism is increasing. More and more tours are being organized for medicinal treatments, including tours to hot springs or national herbal production sites.

School Science Activities

The main mission of the dissemination of science knowledge to the public is to educate through the education system, covering compulsory education from primary/elementary to secondary levels. This also involves higher education in universities or vocational institutions. The dissemination of scientific knowledge in the education system should be conducted in line with the development of appropriate modern curricula. It should emphasize scientific principles and facts, the application of science to beneficial effects and the impacts on society (Schellinger et al., 2017).

Science and technology education must begin at elementary/primary school level so as to set a solid foundation to facilitate learning in secondary/high school (King et al., 2001). Observation, exploration, problem-solving and practical participation are the principles through which to achieve maximum learning competency by combining knowledge in various fields (Raja et al., 2016).

Many elementary/primary schoolteachers lack confidence in science and mathematics. As a matter of fact, they do not always have opportunities to attend any training courses. Governments must promote the understanding of science through appropriate knowledge and school-level curricula and training for teachers at an elementary/primary level.

Science Clubs and science Corners in the School

In addition to the science education provided by governments in accordance with modern society, schools should create an atmosphere in which to stimulate students' enthusiasm for learning science. This may involve activities outside the classroom, such as science clubs for students interested in science to enable them to compare experiences with friends and apply the knowledge they acquire from classroom to club activities.

Schools may reserve a space within their building to be a science corner for science exhibitions created by students on topics of interest, or those related to current issues. In some countries, school science corners are located in special rooms and named STEM Centers.

A science exhibition space may be set in a corner of the school library. The library should be well-equipped with an IT system enabling students to search for information on the Internet, seek knowledge through distance learning and make use of complementary accessories via social media. This will contribute to the expansion of education opportunities and the promotion of further education in different fields.

Science Projects

Science projects in schools are the best way to take students out of the classroom into real life to have fun learning science, technology, engineering and mathematics (STEM) subjects (Mebert et al., 2020). High school students can do research on topics of interest, based on scientific processes learned in the classroom, and present a report of the result to their science teachers or scientists. Moreover, students can prove their initiative in creating their own products by using materials around them. Thus, science project-based learning can contribute to the training of students to have a scientific way of thinking and further help develop the abilities of the talented students to become researchers in the future.

Organization of science project competitions is a way to give a platform to students and encourage them to create and present their scientific works. This platform will also enable students to share and exchange knowledge and experience with other participants working on other projects. Many science museums or science centers host science project competitions, both at national and international level, to promote science knowledge dissemination.

Academic Olympiads are another type of competition that help support students in learning improvement and self-development. Thanks to these competitions, outstanding and talented students likely to become national science leaders in the future can be nurtured.

Science Camp

Science camp is an activity aiming to promote science learning along with the learning of rules and etiquette in living and collaborating with others, through joint activities. The main objective is to allow students to have fun learning science outside the classroom. Activities in physical education can be programmed to ensure the physical health of students. Each science camp is organized around a specific theme, e.g., environmental camps, science toy camps, physics camps, chemistry camps, and so on.

Science plays an important role in social change, which takes place in a dynamic way. Scientific advances and new knowledge of science help to shape a society that lives in a better way. Scientific progress develops at a rapid rate. Therefore, the process of disseminating scientific knowledge to achieve scientific popularity should be undertaken in such a way as to keep pace with social change and scientific progress. The unprecedented COVID-19 pandemic has proved that it is imperative to prepare people to face and solve, through science, the problems caused by this pandemic. We must disseminate knowledge of science widely to people; that is to say, promote an education for an unknown future.

Summary

In this chapter, I have discussed the following: the concept of public understanding of science and technology; science popularization through public lectures and talk shows, science museums or science centers and scientific tourism; and school science activities through science clubs and science corners, science projects, as well as science camps.

Further Reading

Danilina, Y. V. (2022). Science popularization as an element of innovative communications. *Scientific and Technical Information Processing, 49*, 21–29.

Georgopoulou, P., Koliopoulous, D., & Meunier, A. (2021). The dissemination of elements of scientific knowledge in archaeological museums in Greece: Socio-cultural, epistemological and communication/education aspects. *Scientific Culture, 7*(1), 31–44.

Gunay, B. (2012). Musuem concept from past to present and importance of museums as centers of art education. *Procedia Social and Behavioral Sciences, 55*(2012), 1250–1258.

King, K., Shumow, L., & Lietz, S. (2001). Science education in an urban elementary school: Case studies of teacher beliefs and classroom practices. *Science Education, 85*, 89–110.

Mebert, L., Barnes, R., Dalley, J., Gawarecki, L., Ghazi-Nezmi, F., Shafer, G., Slater, J., & Yezbick, E. (2020). Fostering student engagement through a real-world, collaborative project across disciplines and institutions. *Higher Education Pedagogies, 5*, 30–51. https://doi.org/10.1080/23752696.2020.1750306

Ong, E. T., Luo, X., Yuan, J., & Yingprayoon, J. (2020). The effectiveness of a professional development program on the use of STEM-based 5E inquiry learning model for science teachers in China. *Science Education International, 31*(2), 179–184.

Raja, A., Lavin, E. S., Gali, T., & Donovan, K. (2016). Science alive!: Connecting with elementary students through science exploration. *Journal in Microbiology Biology Education, 17*(2), 275–281. https://doi.org/10.1128/jmbe.v17i2.1074

Schellinger, J., Mendenhall, A., Alemanne, N. D., Southerland, S. A., Sampson, V., Douglas, I., Kazmer, M. M., & Marty, P. F. (2017). Doing science in elementary Schhol: Using digital technology to Foster the development of elementary students' understandings of scientific inquiry. *EURASIA Journal of Mathematcis Science and Technology Education, 13*(8), 4634–4648.

Sugimoto, C. R., Thelwall, M., Lariviere, V., Tsou, A., Mongeon, P., & Macaluso, B. (2013). Scientists popularzing science: Characteristics and impact of TED talk presenters. *PLoS One, 8*(4), e62403. https://doi.org/10.1371/journal.pone.0062403

Yingprayoon, J. (2004). *International luncheon on teaching science creatively using locally produced low-cost high-tech materials.* NSTA Convention.

Yingprayoon, J. (2015). Teaching mathematics using augmented reality. In *Proceedings of the 20th Asian Technology Conference in Mathematics,* Leshan, China, December 16–20.

Yingprayoon, J. (2016). Creative mathematics hands-on activities in the classroom. In *Proceedings of the 13th International Congress on Mathematical Education* (pp. 759–760).

Yingprayoon, J. (2018). Students' confidence in participating in the talented teachers of science and mathematics project. In *Proceedings of the 2018 ICBTS International Multidiscipline Research Conference.* ISBN: 978-616-406-885-9 (e-book).

Yingprayoon, J. (2019a). Creativity development in science and technology education. In *Proceedings of MSCEIS 2019, Bandung, Indonesia, October 12.*

Yingprayoon, J. (2019b). *Scientific tourism management in Thailand, actual economy: Local solutions for global challenges* (pp. 304–309).

Yingprayoon, J., Latipov, Z. A., & Sabirova, F. M. (2014). Research on the contribution of the Nobel prize laureates in physics to the development of modern equipment and technologies in technical universities. *European Journal of Science and Theology, 10*(6), 193–202.

Dr. Janchai Yingprayoon received a German Government Scholarship to study a PhD in Laser Physics from Free University of Berlin, Germany. Involved in ICASE since 1979, he became an ICASE Regional Representative in 1993, and later served as ICASE President from 2004–2007. He received an outstanding university lecturer award from the King of Thailand in 1982 and an ICASE Distinguished Service Award at the ICASE World STE2013 Conference, Malaysia in 2013. He has been working as an adjunct professor at Guangxi Normal University in Guilin, China for several years. He has worked extensively with UNESCO and is a creative and humorous lecturer who makes learning science fun and meaningful. He is a well-known international speaker and has been invited to many international conferences as an honorable keynote speaker. Janchai has conducted innovative workshops in 28 countries around the world. He is currently working as a Deputy Director of Demonstration School of Thaksin University, Phatthalung Province, Thailand.

Part III
The Learning Sciences and Twenty-First Century Skills

Chapter 14
Educational Psychology

Keith S. Taber

Abstract This chapter provides a brief survey of key areas of work in educational psychology that are relevant to science and technology education. The chapter offers an introduction, accessible to general professional readers such as teachers, which seeks to: highlight the relevance and value of educational psychology for those working in education; indicate the breadth of the field and some of the key concepts and theoretical areas; introduce key areas of research into teaching informed by psychological perspectives; and encourage readers to consider exploring some of these topics in more depth. The diverse nature of perspectives and methodologies adopted in psychology, and the challenge of studying mental events, are highlighted. The preponderance of informal references to mental phenomena in everyday discourse (the 'mental register') is acknowledged as tending to make some psychological constructs seem more directly accessible and widely understood than is actually the case. Among the topics considered are 'theory of mind', perception, memory, cognitive development, scaffolding learning, metacognition, intelligence, giftedness, motivation and individual differences.

Keywords Cognitive development · Confirmation bias · Gestalts · Giftedness · Individual differences · Intelligence · Meaningful learning · Memory · Metacognition · Motivation · Multi-modal teaching · Neurodiversity · Perception · Self-efficacy · Scaffolding · Theory of mind · Working memory

Psychology Applied to Education

Psychology is recognised to be a diverse discipline, with a broad range of theoretical perspectives and methodologies. *Educational* psychology therefore encompasses *those aspects of psychology applied to examine, enquire into, explain or*

K. S. Taber (✉)
Faculty of Education, University of Cambridge, Cambridge, UK
e-mail: kst24@cam.ac.uk

© The Author(s), under exclusive license to Springer Nature
Switzerland AG 2023
B. Akpan et al. (eds.), *Contemporary Issues in Science and Technology Education*, Contemporary Trends and Issues in Science Education 56,
https://doi.org/10.1007/978-3-031-24259-5_14

193

inform educational phenomena – where education is centrally about teaching, learning and related concepts – curriculum, schooling, and so forth. This comprises a vast body of work, and this chapter can only offer a very brief taster to introduce key areas of work in educational psychology (see Fig. 14.1). A good many of the research studies published in science and technology education draw upon educational psychology for their theoretical perspectives, so some familiarity with this field is important for those seeking to make science and technology education more 'research-informed'.

The chapter format only allows a very brief survey of an entire field of research, intended to be accessible to the general professional reader (such as teachers and others working in education). It has therefore not been possible to cite many original sources: however, the suggestions for further reading provide useful starting points for readers who wish to explore topics further. The chapter therefore aims to (i) highlight the relevance and value of educational psychology for those working in education, such as teachers; (ii) give some indication of the breadth of the field, and some of the key concepts and theoretical areas; (iii) offer a 'primer' for those reading research studies into teaching informed by psychological perspectives; and (iv) encourage readers to consider exploring some of these topics in more detail.

What Is Psychology?

Psychology is the study of the mind. The mind is unobservable, and is indeed an abstract notion that helps us understand people's behaviour. Traditionally, psychology made a distinction between cognition (thinking), affect (feelings), and conation

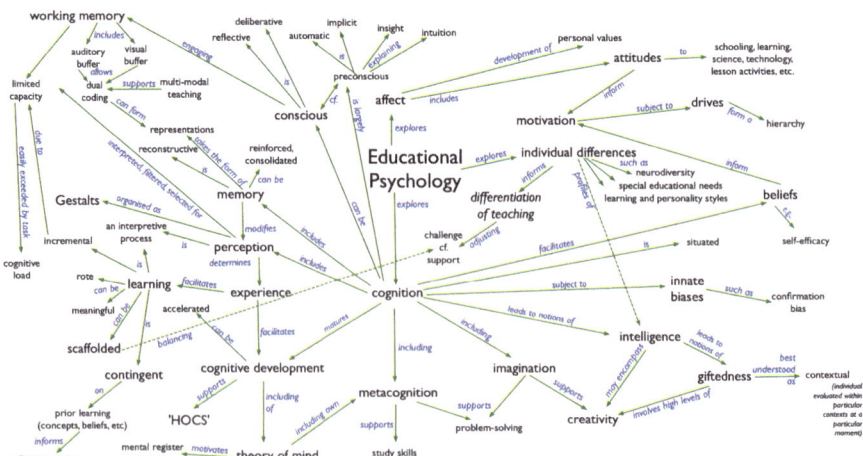

Fig. 14.1 Some of the key ideas from educational psychology relevant to teaching science and technology

(inclination towards behaviour), although such a model may sometimes divide what may be better understood holistically.

Psychology is a broad discipline that has encompassed a range of quite different traditions. For example, most people know something about the psychoanalytic notions of theorists such as Sigmund Freud, Carl Jung and others. Famously, Freud would seek to interpret his clients' dreams in terms of their unconscious fears and desires. The psychoanalytic tradition is still active today, even if it is not commonly used in educational research. But much educational psychology does use interpretive methods to make sense of data (for example, from detailed interviews). Yet, it is also the case that other psychological research follows 'paradigms' (widely used outline research designs) adopting experimental methods much more akin to the kinds of investigations familiar from the school physics laboratory – controlling variables, compiling tables of quantitative results, and using statistical methods to reach conclusions.

One influential school of psychology, behaviourism, was informed by a perspective that sought to exclude unobservable entities from psychological explanations. That was in contrast to a tradition that admitted introspection (reflecting on one's own mental experience) as a useful tool. Although behaviourism is no longer a leading approach, the relative value of what can be objectively measured as opposed to subjective reports of personal experience remains a key issue.

As an extreme example, positron emission tomography (PET) scanning, an objective technique, may be used to image a person's brain as they carry out some kind of cognitive task, but cannot be *directly* linked to mental experience, whereas a talk-aloud protocol can allow a study participant to describe their conscious thinking processes – but only offers a subjective account subject to biases and selective reporting, and requiring the analyst to interpret the participant's intended meanings.

Science and technology teachers may tend to be more convinced by research methods that *seem* more objective. Quantifiable entities may appear more 'real' than those relying on the analysts' interpretations, yet many constructs 'measured' in educational psychology (e.g. measuring *attitude to science*, measuring *self-efficacy*, measuring *science capital*) depend on Likert-type scales (e.g. 'please rate your agreement on a 6 point scale where 1 is strong disagreement and 6 is strong agreement') made up of sets of statements that have been found to be statistically associated among samples of respondents, and where individuals are asked to subjectively score at one moment in time.

Applying experimental methods to investigate teaching and learning is very challenging (Taber, 2019). For example, when working with human participants, there are serious complications that can confound experimental results (e.g., it has been widely demonstrated that a participant's *expectations* about what an experimental treatment might achieve – *which may be inadvertently communicated by researchers* – can strongly influence outcomes). Often educational research is undertaken with non-random samples from populations where it is simply assumed that '11–12-year-old schoolchildren' or 'teenagers with a diagnosis of autism' can be considered as if natural kinds, where 'specimens' are interchangeable when using statistical methods to reach conclusions. However, research actually shows that study findings may not unproblematically transfer between very different educational contexts.

The Mental Register and Theory of Mind

Psychologists have studied human development, and found patterns in areas such as developing sophistication of thinking and committing to a coherent set of personal values. Generally, human beings naturally develop what is known as 'theory of mind' (TOM): from quite an early age, we learn to automatically consider what others around us will think about what we do and speak. This means that we are all inherently *implicit* psychologists and it is very common for everyday discourse to include references to mental entities and events.

Whilst our mental experiences relate to consciousness, research suggests that we are only aware of a fraction of our cognitive processing – most is 'preconscious'. That so much of our 'thinking' is tacit becomes clear when we consider intuition (feeling that we know something, without having any idea of how we came to know) or when we have a sudden insight regarding some problem or issue (Brock, 2015).

There is a 'mental register' (cf. Box 14.1) of terms that we commonly use in relation to mental phenomena – remembering, learning, intelligence, understanding, knowing, thinking, and so forth – but when these terms are used in everyday discourse, they have somewhat vague meanings (Taber, 2013). In professional (e.g., educational) discourse, terms that might be considered to have a technical meaning need to be more clearly (i.e., operationally) defined, and educational psychology research often provides such clarity.

> **Box 14.1** The mental register: a selection of words relating to mental experience (Taber, 2013). Some of these words (a) are largely restricted to use in everyday discourse; but some (b) are also commonly part of the professional discourse of those working in education; and some (c) are used as technical terms in academic psychology and educational research. There is a potential for confusion where terms that may have *a well-defined technical meaning in research are also used in more fluid and diffuse ways within professional educational discourse*
> belief, clever, cognition, comprehension, concentrating, contemplation, creativity, cunning, day-dreaming, distraction, evaluating, focusing, forgetting, gifted, idea, imagining/imagination, ingenuity, intelligence, insight, intuition, knowing/knowledge, learning, memory, mental imagery, mind's eye, misconception, metacognition, perception, problem-solving, rational, reasoning, reflection, remembering, reverie, self-knowledge, thinking/thought, understanding, wisdom...

Perception

We might understand perception as the means by which we acquire data about the outside world – we see, hear, touch, and so forth. In education, perception provides experiences from which a person can learn. The student looks at the textbook or board or screen, hears the teacher, feels how the test tube has warmed-up, smells the vapour above the ammonium hydroxide solution, and so forth. Understanding how our sensory organs function is part of *physiology*. However, *psychology* has explored perception within a systems perspective – not how neural signals are generated (for example, in the retina) and transmitted, but how those signals are *interpreted* within the mind of the individual.

Making sense of perceptual data depends on innate and learned biases. Confirmation bias leads to more readily noticing evidence supporting rather than challenging beliefs – so teachers have to remember to draw attention to, and explicitly explain, evidence intended to challenge learners' ideas. Chicks will peck at certain simple abstract shapes that look (to us) nothing like parent birds, because the chicks are primed to respond to simple cues when being fed: similarly, humans seem to have an innate predisposition for making sense of minimal data to see faces:

(-: 😁).

Most people are familiar with optical illusions, which show that perception is not just a matter of our senses telling us what is in the world. So, for example, there are ambiguous figures that can be seen in two ways (e.g., is it a rabbit, or a duck?) that were studied by the 'Gestalt' psychologists who explored how the way humans experience the world has to be understood as more holistic than simply treating perceptual data as if pieces of some jigsaw puzzle that can be put together to give an authentic picture of the world. For example, Fig. 14.2 is an image that can be seen in two different ways.

The figure might be seen as two faces (shown in white against a black background). Or the same image (i.e., the same perceptual *data*) could be seen – that is interpreted – as some kind of goblet or candlestick holder (in black against a white background). A person can learn to see either version, but not both at the same time. The brain actively organises perception to make sense of the image, and the viewer can force the 'Gestalt-shift' between the two interpretations. Moreover, during the shift you can 'see' the goblet emerging from the background (or *vice versa*), giving an *impression* of depth.

So, even though I know I am looking at a flat image, I can see the goblet 'move' into the foreground. Perceptual organisation takes place at a preconscious level, not under our deliberate control. Once one has 'seen' a face in cloud formations or in a pattern of craters on the moon, it cannot be simply 'unseen' because *we know* it is just an artefact.

Teachers need to appreciate that perception depends on the brain being primed by prior experience. The teacher might draw or present an image that, to the teacher,

Fig. 14.2 An ambiguous representation. (Image by ElisaRiva from Pixabay)

clearly shows some experimental set-up, plant structure, or three-dimensional configuration of a molecule. But the teacher cannot *assume* that the learners in the class see the same thing. Seeing relies on preconscious brain programming that has been cued by earlier experiences. There are stories of indigenous people being shown photographs or drawings for the first time, but making no sense of them – whilst those same people immediately appreciate the symbolism of their own, familiar, cultural artefacts. Scientists learn to see structures through the microscope or to spot fossil fragments in the field that the untrained observer simply does not see.

Equivalent perceptual data can therefore be organised differently inside different people's brains. Not only does the teacher need to check the learners' interpretations, but she may need to patiently teach the learners a new way of seeing.

Memory

Memory is something that we may tend to *assume* we understand: we 'know' that we store information in memory, and later remember it, but sometimes we forget. As with perception, research shows that the way we talk about memory in everyday discourse is a considerable simplification. We can better think of memory as being some trace of past experience that can potentially change our current behaviour. Educational psychology has produced important findings about how the extent of remembering or forgetting information depends on the number and timing of repeat exposures – something very relevant to both teachers and revising students.

More importantly, research suggests that the naive TOM model of memory is not accurate in two important ways (Taber, 2013).

Firstly, memory does not work like 'storage' (where we can enter the store and retrieve the original object), but as forming 'representations' that have to be interpreted later. Memory is 'reconstructive', as those representations are the basis for the preconscious building of a 'memory' that makes sense within our wider understanding. Research shows that memories are not simply present, absent or incomplete, but often quite distorted compared with the original experience that was represented – 'eye witness' testimony has been found to be often unreliable, even when a witness is trying to be completely honest about what they saw.

One study about teaching electrical concepts showed that even when learners' alternative conceptions seemed to be successfully challenged by critical laboratory demonstrations, some weeks later the students were likely to remember having seen the demonstrations as supporting their original deeply-held alternative conceptions (Gauld, 1989). The details of what they had seen were (inadvertently) incorrectly remembered in a way that made better sense to the students.

This is very important for teachers as it shows that learning cannot be assumed to occur by a simple accretion of knowledge, concept by concept, lesson by lesson. Memories are consolidated over time, and are initially fragile (open to becoming inaccessible or being distorted) until there has been sufficient reinforcement. Key ideas need to be revisited throughout the course, so the curriculum has to be designed accordingly.

The second counter-intuitive feature of memory is that it is not really a discrete faculty of mind. What we learn does not just change what we can try to consciously access, but also modifies the functioning of the parts of the brain that filter and interpret perceptual information. We do actually *learn to see* (and hear, etc.) differently through experience.

Cognitive Development

Cognition is a blanket term referring to those (preconscious and conscious) processes that support a person in learning and thinking about the world – developing knowledge and understanding. There is much work on various aspects of cognition that is relevant to teaching. A common tool used in planning teaching is to distinguish so-called LOCS and HOCS (lower-order and higher-order cognitive skills). In general, asking a student to undertake tasks such as recalling and applying previously learned information is of lower demand than asking them to analyse, critique or evaluate material, or to solve problems or create new products. Planning teaching needs to be informed by awareness of the stage of development and prior knowledge and skills of the learners; but supporting *development* requires learners to be challenged – which means including tasks requiring HOCS.

There has been much work on problem-solving (Tsapalis, 2021), which is the ability to succeed in tasks that are novel (see Chap. 18). (So, *finding a method* for solving a quadratic equation would be solving a problem, but *applying a previously learned rule or algorithm* would not – even though the outcome is the same.) The very limited capacity of what is known as 'working memory' is critical in determining what learners can achieve. A task that overloads working memory is very likely to lead to failure. Sometimes, people can learn strategies and techniques that can work around these limitations, and learning about this area of research can help teachers to design tasks and assessment items that do not inherently lead to most students failing. Whilst substantive learning depends on learners being challenged, success also depends on providing sufficient support to engage with the challenges. The concept of scaffolding is especially important here, as it suggests how learners can be supported to take on tasks beyond their current competence (important to encourage substantive development), but only with the right kind of structuring and support that can be 'faded' away as the learner develops new competences. Achieving the correct challenge/support balance can be key in facilitating 'flow' experiences – where engagement is such that there is intense concentration and students 'lose themselves' in activities.

There is also research on human thinking that shows that some of the reasoning and logic that is taken for granted in science and technological fields cannot be assumed to be available to young learners. Most famously, the developmental psychologist Jean Piaget characterised aspects of children's thinking at different ages, and identified the kinds of thinking available to children at various stages in their schooling.

Although Piaget's ideas are no longer *fully* accepted, this is in part because further research building on his findings has identified ways of presenting information and tasks to support learners in achieving tasks earlier than would otherwise have been possible (Bliss, 1995). Luria's work has suggested that education influences the forms of thinking available (or at least, usually employed) in a culture. The 'CASE' (*Cognitive Acceleration in Science Education*) project developed tasks to prepare students entering adolescence to master the kinds of thinking needed to engage with the abstract, theoretical side of science (Adey, 1999).

This project also built on the work of the Russian researcher Lev Vygotsky, whose research led to the notion of scaffolding learning – how, in the right social context, students can be supported to develop new skills and thinking by engagement in shared structured activities where they incrementally take on more responsibility for activity. Vygotsky also offered models for helping us understand how a person's knowledge and understanding is built by the *interaction* of direct experience and formal symbolic instruction (that is by doing, *and* also by reading or being told). This has direct relevance for teachers charged with helping learners understand abstract concepts that are not directly linked to what students can immediately experience (that is, much that is taught in science classes).

This area of work is especially relevant for science teachers, who are often charged with presenting highly abstract content that cannot be *directly* demonstrated. This may require learners to undertake mental operations on abstract ideas – just the kind of mentipulation that Piaget suggested only slowly developed during adolescence. Yet, there is now much work showing that teaching approaches can often be found to overcome this. A key notion is that of the educational psychologist David Ausubel (1968), who suggested that the most important factor in a student's learning is what she already knows. The constructivist account of learning has made it clear that, whilst Piaget was right to assume that conceptual development has *its foundations* in direct experience of the material world, that which starts off as abstraction will, with sufficient familiarity, become *as if* concrete over time (so, for a science teacher, a methane molecule, or a food web, or a ray of light has become *as familiar* as any concrete object and can be readily mentipulated). Ausubel referred to *meaningful* learning (contrasted with rote learning), where the learner makes sense of teaching because it is associated with existing knowledge. Teachers can seek to introduce new abstract ideas by showing that they can be understood in terms of the conceptual resources (ideas, images, experiences, etc.) that their students already have available – 'making the unfamiliar familiar' through the common use of metaphor, simile, analogy, personification, anthropomorphism, narrative, etcetera, in teaching.

One important area of research on cognition emphasises how it is often embedded within a social context. This work suggests that we cannot assume that we can transpose the same individual to a different context (perhaps removing them from a context where they have developed and demonstrated some competence) without influencing their ('situated') cognitive processing (Hennessy, 1993). Context is also important in an 'internal' sense: recent thinking may provide a 'set' that is likely to channel current thinking in a particular way. (This may be of special significance in schooling, where very often the learner in front of the teacher has just come from learning about a completely different discipline.)

Metacognition

Research has also shown the importance of metacognition in higher level learning – that is the ability to be aware of, monitor and direct one's own learning processes – to be able to make one's own thinking the subject of reflection. In a classroom learning situation, the teacher cannot be directly monitoring all the learners at once, so there is great value in a student knowing when they need help, when a change of activity or a break would be more productive than just continuing with an activity, or when they have sufficiently mastered a task and are ready to move on.

Metacognition allows students to be self-directed learners (so, for example, revision is not just endless re-reading of notes) and to take on projects that need to be planned, monitored and evaluated. Students can be supported in using and developing metacognitive skills, and teachers can include 'metacognitive prompts' in instructional materials to remind students to periodically step back from their direct engagement in tasks, to engage in a metacognitive review (Zohar & Barzilai, 2013).

A useful finding for teachers to be aware of is the Dunning-Kruger effect, whereby the least able students tend to over-estimate their achievements (so that they tend to expect that they will do better in a test or exam than proves to be the case). This can be seen as linked to the concept of self-efficacy, which concerns a person's belief in their ability to successfully attain certain goals. Whilst this is an individual characteristic, there may be cultural influences that lead to group effects. For example, students who are aware that they are in a 'bottom set' may believe they have little ability in a subject and cannot achieve anything without continuous step-by-step instruction (despite work being set to match their development and competence) and, certainly historically (and perhaps even today), there has been a tendency for girls to have less self-efficacy in mathematics and physics.

Intelligence

Intelligence is perhaps a classic example of a concept that is widely used in public discourse, but where there is no single, precise, agreed meaning. Most people are aware of IQ, intelligence quotient, as a score obtained from a kind of intelligence test. Testing was first introduced as a means to identify learners in school classes who were not able to benefit (those whom might sometimes be seen today as children with special educational needs), and needed to be separated out for different instruction.

Despite this valuable aim, IQ tests were at one time used to class the population into ability ranges, using terms such as 'idiot', 'imbecile' and 'moron', which came to be adopted as terms of derision. Although IQ was found to be a fairly reliable measure (individuals tend to have fairly stable scores), early IQ tests were found to be culturally biased. They included questions that would only be understood by people sharing particular cultural knowledge (such as how the playing field is laid out in baseball).

It was also found that, over time, the scoring of IQ tests had to be adjusted to avoid the average population score drifting upwards – an indication that IQ scores do not capture something entirely innate, but have reflected global improvements in educational provision. It is sometimes considered that intelligence reflects two components, one reflecting biological features (e.g., how quickly signals are transmitted along nerves) and the other based on an individual's past experience, and so learning.

Intelligence has been modelled in various ways. One approach distinguishes a generalised factor (g), which is characteristic of an individual, but which is moderated in different domains depending on the individual's relative expertise. So, in this model, one would only become a brilliant microbiologist by *both* having a high 'g' *and* by committing time to develop expertise in microbiology. The general factor is also sometimes divided into a 'crystallised' factor (related to applying *prior* learning) and a 'fluid' factor (supporting problem-solving). The theory of multiple intelligences, however, suggests that it is better to consider intelligences as a set of largely discrete characteristics of an individual. Gardner's (1993) model includes: linguistic, logical/mathematical, spatial, bodily-kinaesthetic, musical, interpersonal, intrapersonal, and naturalist intelligences.

According to this model, the IQ test is an incomplete measure of intelligence, as it primarily tests only a subset of these largely independent intelligences (excluding, for example, the 'naturalist' intelligence highly relevant to science learning). The degree of 'modularity of mind' (comprising of largely discrete components, rather than general purpose abilities) remains a debated question. Some other models of intelligence, such as those developed by Sternberg, also downplay IQ. He considers intelligence to be the ability of someone to achieve personally meaningful goals, drawing upon four areas – creativity, analytical ability, practical ability and wisdom – but also by being aware of best utilising personal strengths and weaknesses (i.e., applying metacognition). Teachers should be aware that there is more than one way to define and measure intelligence.

Learning Styles

A learning style is a *preferred* way of learning. There are many models of learning styles, but most are only supported by weak empirical evidence that they reflect genuine stable differences between individuals. A very common notion that has been mooted as 'learning style' is known as VAK, suggesting that learners tend to vary on a profile as primarily Visual, Auditory or Kinaesthetic learners. Although very popular, this seems to be largely a misuse of the multiple intelligence concept, and there is no strong basis for seeing these modalities as learning styles.

However, multi-modal teaching (that uses words, images, gesture and practical activity) is potentially very useful, as information received via the different senses can be mutually reinforcing. This is particularly so when images complement spoken information – as long as learners can see how the two are related. Research has led to a model of working memory that includes two small data buffers, one of which can temporarily store a small amount of data originating from the visual system, and the other able to store a small amount of data deriving from the auditory system, as well as a 'space' to mentipulate this data. This potentially allows what is known as 'dual coding', forming associated representations of verbal and imagistic information: something that is considered to aid later recall. As with many such theories in psychology, the entities discussed (the working memory components, the representations) cannot be directly observed, and so evidence is indirect and different interpretations of the evidence are possible. (The same can, of course, be said of many theoretical entities in the *natural* sciences: the Higgs boson, dark matter, mitochondrial Eve, and so forth.)

Gifted Learners and Differentiation

It has been common in some educational contexts to seek to identify learners who are 'high ability', 'exceptional', 'gifted' or 'talented'. These attempts have usually been well-intentioned, as all students are entitled to educative schooling and students who are more advanced in their learning may sometimes not be sufficiently challenged in standard classes in order to support their development. There is no agreed definition as to what counts as 'gifted', or how to identify the 'gifted' (sometimes IQ tests are used, sometimes teacher recommendations, or diagnostic checklists). Dividing learners into 'gifted' and 'others' (so, not gifted) may be divisive, and may become a self-fulfilling prophecy, as well as a cause of resentment. Also, labelling individuals as 'gifted' may put them under stress if they feel excessive expectations.

Giftedness is better seen as *contextual*, as the student who is seen as gifted in mathematics may not show special abilities in, say, biology, and even within a discipline students have different strengths and weaknesses. Teachers should look to differentiate work such that all students in a class are asked to undertake tasks that they consider challenging, but for which sufficient support is provided to allow them to succeed. This need not always mean differentiating by offering different tasks, but could be differentiating in terms of level of support, or even role. For example, some students may sometimes be asked to act as mentors or tutors to other students; providing that they are comfortable doing so, they are offered support in developing skills in the role, and this is designed as a learning opportunity for the mentor as well as their peers (as preparing for teaching and developing learning materials can be effective and demanding learning activities).

Affect and Learner Motivation

Whilst science and technology teaching often focuses on cognition – developing knowledge and understanding, problem-solving and the like – the aims of teaching also encompass affective (and aesthetic) values. We want students to develop certain attitudes to the natural world, and to science and technology. Research suggests that just as young people slowly develop more sophisticated cognition during their school careers, they also develop nuanced personal systems of values.

Attitudes are extremely important in education (Potvin & Hasni, 2014). For example, a student who does not value schooling, or does not value a particular school subject, is unlikely to be motivated to commit to a high level of effort. Such attitudes may derive from home values (students from certain backgrounds are given an advantage in education because of the values and attitudes habitually expressed in the home environment).

This may link to personal belief systems about whether formal qualifications are important for adult life, and also to personal beliefs about the self: whether a student thinks of herself as a good student (capable, clever, able, etc.); whether she considers that she has potential in particular curriculum areas (historically, in many contexts, there have been subtle, or even blatant, clues suggesting science and technology are areas more suitable for boys than girls). A student from a home where academic qualifications and schooling are not valued, who does not see themselves as academic, and who thinks that they are part of a group unsuited to science, lacking role models in science and technology, and without ready access to science careers, has little reason to have high expectations of, or to commit to being conscientious and industrious in, science learning. In some cultural contexts, science is often assumed to be a more difficult curriculum area than the humanities or social science. Good teachers may challenge such attitudes and beliefs, but need to be aware that these may be well-established, and that extended positive experiences may be needed to bring about long-term change.

One of the key theorists in motivation theory was Maslow, who proposed a hierarchy of human needs in which the individual is driven to meet the most basic needs (such as food and shelter) before they can focus on higher goals. Even if that is an over-simplification, teachers cannot expect students to concentrate on academic learning when they come to school hungry, or are frightened (perhaps being abused at home or bullied in the playground), or feel unloved (perhaps interpreting a parental break-up as their own fault). The teacher should be alert to such possibilities and, when indicated, involve appropriate other agencies.

Individual Differences

Psychology has both identified commonalities among learners (for example, in the general pattern of cognitive development over time) and the importance of individual differences (such as Gardner's notion of profiles of 'intelligences'). One area where the teacher may come into contact with professional educational psychologists is in the identification of learners with special educational needs, such as those who have specific learning difficulties such as dyslexia. Following a tradition that goes back at least to Vygotsky (and indeed Binet's original IQ testing), psychologists have looked to inform teachers about how to best support learning in students with individual differences that may otherwise act as barriers to learning.

In recent years, there has also been increasing attention given to so-called neurodiversity, which is concerned with differences in mental functioning. This covers a range of types of individual differences, including the autism spectrum disorder and indeed such conditions as synaesthesia (where a person experiences somewhat conflated senses). There are many dimensions along which people can vary, and identified groups ('the autistic', 'the gifted', synaesthetes, etc.) may be better understood as those found at the extremes of some of these shared dimensions. That is, neurodiversity refers to ranges of variation on which we can all be located.

Summary

This chapter has introduced some of the wide range of areas in which educational psychology can inform teaching of science and technology. As the examples presented suggest, work in educational psychology links with, and indeed has sometimes motivated, major initiatives in science teaching and STEM curriculum development. Much of the foundational work in educational psychology has, naturally, been carried out by psychologists who see education as a context for research. Whilst applications are often suggested by studies in this field, it is often left to those more centrally working in education to explore implications and develop practical interventions based on psychological research. This often means that educational researchers, teacher developers and curriculum developers in science and technology need to build on the foundational work to develop implementations for classroom practice.

Recommended Resources

The following volume includes chapters on a range of educational theories and theorists, including several important perspectives from educational psychology: *Science Education in theory and practice: An introductory guide to learning theory*, Akpan, B. & Kennedy, T. (Eds.). (2020). Springer.

A book that develops many of the ideas in this chapter, and offers specific examples of how principles from educational psychology can be applied in science teaching is: *Masterclass in science education: Transforming teaching and learning*, Taber, K. S. (2018). Bloomsbury.

An edited volume looking at affect in science education is *Affective dimensions in chemistry education*, Kahveci, M. & Orgill, M. (Eds.). (2015). Springer.

Various approaches to meeting the needs of gifted learners are discussed in *Policy and Practice in science education for the gifted: Approaches from diverse national contexts*, Sumida, M. & Taber, K. S. (Eds.). (2017). Routledge.

Examples of approaches to 'make the unfamiliar familiar' to link science concepts with students' existing conceptual resources can be found at: https://science-education-research.com/teaching-science/constructivist-pedagogy/making-the-unfamiliar-familiar/

References

Adey, P. (1999). *The science of thinking, and science for thinking: A description of cognitive acceleration through science education (CASE)*. International Bureau of Education (UNESCO).

Ausubel, D. P. (1968). *Educational psychology: A cognitive view*. Holt, Rinehart & Winston.

Bliss, J. (1995). Piaget and after: The case of learning science. *Studies in Science Education, 25,* 139–172.

Brock, R. (2015). Intuition and insight: Two concepts that illuminate the tacit in science education. *Studies in Science Education, 51*(2), 127–167. https://doi.org/10.1080/03057267.2015.1049843

Gardner, H. (1993). *Frames of mind: The theory of multiple intelligences* (2nd ed.). Fontana.

Gauld, C. (1989). A study of pupils' responses to empirical evidence. In R. Millar (Ed.), *Doing science: Images of science in science education* (pp. 62–82). The Falmer Press.

Hennessy, S. (1993). Situated cognition and cognitive apprenticeship: Implications for classroom learning. *Studies in Science Education, 22,* 1–41.

Potvin, P., & Hasni, A. (2014). Interest, motivation and attitude towards science and technology at K-12 levels: A systematic review of 12 years of educational research. *Studies in Science Education, 50*(1), 85–129. https://doi.org/10.1080/03057267.2014.881626

Taber, K. S. (2013). *Modelling learners and learning in science education: Developing representations of concepts, conceptual structure and conceptual change to inform teaching and research.* Springer.

Taber, K. S. (2019). Experimental research into teaching innovations: Responding to methodological and ethical challenges. *Studies in Science Education, 55*(1), 69–119. https://doi.org/10.1080/03057267.2019.1658058

Tsapalis, G. (Ed.). (2021). *Problems and problem solving in chemistry education. Analysing data, looking for patterns and making deductions.* Royal Society of Chemistry.

Zohar, A., & Barzilai, S. (2013). A review of research on metacognition in science education: Current and future directions. *Studies in Science Education, 49*(2), 121–169. https://doi.org/10.1080/03057267.2013.847261

Keith S. Taber is Emeritus Professor of Science Education at the University of Cambridge. He taught science in schools and further education, before joining the Faculty of Education at Cambridge, working in initial teacher education, lecturing in research methods, and supervising graduate research projects.

Chapter 15
Science and Technology Teaching Strategies

César Mora

Abstract We present a brief selection and discussion of the main and most successful teaching methodologies for science and technology nowadays. Among them, we have the Active Learning of science and technology, Interactive Lecture Demonstrations and the Project Based Learning Method, the STEM model, and the inquiry method learning methodology. Also, the problem-solving methodology and the use of interactive computational simulations; as well the laboratory as a teaching strategy and History and philosophy of science in teaching.

Keywords Science and technology education · Didactic of science · Teaching strategies

Introduction

At present, science education and technology education are disciplines that bring together different proposals for models and methodologies developed over the years in order to improve the teaching and learning of science and technology at all educational levels (Callahan & Dopico, 2016). For 50 years, ICASE has promoted various regional activities across the world to improve the teaching of science and technology. However, as Voelker et al. (2006) pointed out, we cannot say that science education is a trivial problem, in which it is enough to use some successful methodology in order to obtain the best learning results after its application, since the educational problem is very complex and has great social implications, especially when concepts associated with mathematical language are involved (Yeo & Gilbert, 2022). Therefore, what is the best methodology for teaching science and technology? Should some methodologies be discarded because they are outdated?

C. Mora (✉)
Centro de Investigación en Ciencia Aplicada y Tecnología Avanzada Unidad Legaria, Instituto Politécnico Nacional, Mexico City, Mexico
e-mail: cmoral@ipn.mx

© The Author(s), under exclusive license to Springer Nature Switzerland AG 2023
B. Akpan et al. (eds.), *Contemporary Issues in Science and Technology Education*, Contemporary Trends and Issues in Science Education 56,
https://doi.org/10.1007/978-3-031-24259-5_15

How can we solve the problem of educational research achievements and their non-inclusion in classroom teaching? These are just some of the questions that remain without a definitive answer, and yet they can be approached from a varied perspective, considering the achievements and advances obtained from Active Learning in science, a discipline that is very versatile and easy to apply. In this chapter, we show some of the most successful methodologies of today. We have not focused on making a global count of the best proposals, since space is limited and the main objective is to show what is working in the present.

The content of the chapter is organized as follows: in the next section, we show some rudiments of Active Learning of science and technology, and briefly describe the Interactive Lecture Demonstrations, as well as the Project Method. We then address the description of the STEM model of teaching, and the discovery learning methodology. In the following section, we present the problem-solving methodology and and the interactive computational simulations, after which the laboratory is mentioned as a teaching strategy. We then go on to consider the importance of the history and philosophy of science in teaching and, finally, we present our conclusions.

Active Learning of Science and Technology

Active Science Learning in its 'hand-on minds-on' version is one of the most successful methodologies in the last twenty years, and this has been emphasized even during the period of the global COVID-19 pandemic at all educational levels, from elementary to university. The different Active Learning approaches *'put students more in the driver's seat through discussion, class questions, and feedback; interactive technologies; and other strategies to enroll learners and deep learning'* (Yannier et al., 2021). There is no single Active Learning approach; rather, there is a wide variety of methods and ideas to produce more effective learning. In what follows we will mention some important features of Active Learning.

The term 'Active Learning' depends on the context and who is using it. On many occasions, it is used interchangeably when talking about Collaborative Learning or Co-operative Learning. It can encompass a variety of activities, including students discussing a problem or concept with another student during class throughout the semester. Active Learning basically means that the students are involved in some kind of guided activity in the class, so that they are doing something in the classroom besides sitting and listening to the instructor give a lecture, or looking at work problems on the classroom blackboard (Meltzer & Thornton, 2012); in this way, in the classroom, students are not passive recipients of knowledge but rather are active learners, and teachers are no longer seen as sources of information, but as moderators or mediators. In Active methodologies, there must be an interactive participation from students to achieve conceptual understanding through hands-on and minds-on activities, which produces immediate information through discussion with their peers and/or instructors. All this can be done based on what is observed in the graphs obtained in real time, or in the analysis of computer simulations, or in the solution of interactive problems, among other options.

Among the activities that are suggested to be done in the Active Learning methodologies are the following:

1. Work in small groups (2 to 4 students) and let the students interact freely, with the teacher more of a moderator or mediator.
2. Encourage students to get to know each other and to participate in different roles, such as secretary, presenter, team leader, etc.
3. Ask students questions during classes to stimulate curiosity.
4. Take a short five-minute test at the beginning of each class.
5. Carry out simple experiments in class and use visual graphics.
6. Encourage critical and independent thinking.
7. Use the Socratic method to ask questions and encourage collaboration among students.
8. Use guides and written reports that help to carry out activities, as well as learning notebooks.
9. Give team reports before the whole group and hold debates to defend the ideas and conclusions.
10. Ask students to evaluate each other's work.
11. Ask students to document their learning progress.
12. Work on completing projects and break them down into smaller parts.
13. Use the discovery learning method and encourage self-assessment.
14. Address problems of everyday life that are significant for students.
15. Use technological resources to control and record data.

This list brings together some recommendations and there may be more suggestions depending on the educational conditions experienced by both students and teachers. However, it should be found that the learning activities follow an order and a well-defined pattern to guarantee learning. For this, we rely on a cognitive cycle that includes predictions, discussions in small groups, observations, and discussions of observed results with the predictions that will allow a synthesis of what has been learned. This cycle is known as the PODS cycle in science education. Sokoloff and Thornton (1997) developed a methodology known as Interactive Demonstrative Classes, which uses the collection of data generated in real time through computer-assisted laboratory tools, and basically consists of the following:

1. The teacher describes the demonstration and performs it without making the measurements.
2. Students are asked to make their predictions in writing and individually,
3. It is then proposed that they work in small groups, with the instructor showing them the most frequent predictions made by the students, so that they then make their final predictions.
4. Later, the instructor carries out the demonstration with measurements shown through graphs produced by the software that is used, and which are presented through a projector.

5. Afterwards, the students describe and analyze the observed results.
6. Finally, the students discuss with the instructor any other similar physical situations upon which the same kind of ideas and concepts can be applied.

Finally, we will mention the Project Method as an active methodology; generally, this method emphasizes the unification of theoretical and practical learning, the collaboration of students and the inclusion of elements of daily life in educational institutions. The method is defined by the following five points:

1. Learning is based on genuine interest and/or initiative.
2. Students discuss their interests and alternative perspectives on the topic, advising one another.
3. They develop their own sphere of activity (limiting proposals, planning, making team decisions, using low-cost materials, etc.).
4. They suspend their activities from time to time, to reflect on their actions, exchange ideas, hold debates, rethink the project, etc.
5. The project ends at a certain point, when the task to be carried out has been achieved.

Due to the characteristics of the Project Method, therefore, there is also a direct connection with the STEM methodology, in such a way that both Active Learning and STEM education use the development of projects as teaching strategies in the same way, although with different approaches.

The STEM Model of Teaching

One of the fastest-growing strategies to teach science and technology today is the STEM model (Li et al., 2020). Certainly, we are facing a new educational paradigm in a new industrial revolution; in this way, the STEM model is a relatively recent term and we can find its origins in the 1990s in the US. At the end of the last century, the main international reports indicated a low performance and interest on the part of American science students, as well as the loss of the country's economic competitiveness due to the lack of qualified professionals to face the emerging context of a new century and millennium. In this way, the STEM model emerges as an educational and economic strategy that aims to facilitate the construction of knowledge and the development of skills considered essential for the context of the global challenges of the twenty-first century.

Initially, the educational tendency to unify science, technology, engineering and mathematics was focused on elementary education and, little by little, the model has been extended to higher education. This integrative model has also been used to include disciplines in psychology, social sciences, and art. Beyond the controversy over whether STEM is a model, a theory, or a strategy, research reports show that

STEM education is one of the main approaches to science and technology education in the last 20 years. Indeed, STEM education can be viewed from a perspective of interdisciplinary or transdisciplinary combinations of the individual STEM disciplines, or also from an inclusive perspective of particular educational disciplines, that is, science education, technology education, engineering education, and mathematics education. All this produces a diversity of approaches in STEM research.

Over time, the STEM movement has gained strength in other countries, including Canada, the UK, Morocco, South Africa, Australia, Germany, Finland, France, Italy, Israel, Japan, China, Turkey and Latin America. However, the movement is still presented in a reduced form, although in an upward growth. In global terms, the celebration of congresses and symposia continues to be scarce and these seek to promote research and implementation of practices focused on the STEM model. After the global COVID-19 pandemic, the use of ICT has increased and, in a certain way, its use has been instrumental in bringing us to education in a digital world. In addition, various STEM scenarios have been disseminated that previously were only available in person. This has facilitated the diffusion of the STEM methodology in Latin America, since, initially, it had been considered a North American methodology and had been used mainly in private elementary schools. In Latin America, the Latin American Science Education Research Association (LASERA) has been one of the main promoters of training and the dissemination of educational work on STEM through the organization of seminars, workshops and conferences.

STEM is inquiry-based science education, an innovative methodology in which the student approaches the concepts through steps to the scientist (Kennedy & Odell, 2022). The objective is to develop, in the student, skills related to a specific job, including the capacity for critical observation and description, both orally and in writing; developing the ability to obtain data and order results in a meaningful way that allows the student to analyze, interpret, establish similarities, differences and, through analysis, reach possible conclusions and hypotheses; and finally, relating the results, predicting others in comparable situations and proposing new experiences to confirm or refute the hypotheses based on experimental evidence. Likewise, through joint work in the interdisciplinary STEM perspective and its application to real problems, it is intended to complement the learning of scientific and technological content by promoting the development of divergent thinking and the increase in student creativity.

One of the main ways in which the STEM methodology has been used for teaching science and technology is through auxiliary educational methodologies, such as project-based teaching, peer instruction, or through Interactive Demonstration Classes, where collaborative learning is applied. These methodologies seek to relate the subject of study with the teaching of science and technology. As well as providing a critique of the subject, with professional guidance giving the pros and cons, it also seeks to relate the topic to current/proposed teaching practices and actual applications in the classroom. Finally, examples of model practices, ideas and/or programs are provided.

We will end this section by mentioning the different skills that we seek to develop in students who use the STEM model. First of all, we must bear in mind that the foundation of STEM education is the integration of disciplines and skills to achieve effective learning of science and technology. Therefore, the student must be able to:

1. Investigate: that is, the inquiry learning methodology is used. Campanario and Moya (1999) mention that discovery learning tends to be associated with primary and secondary education levels and, in fact, it was one of the first alternatives offered to traditional repetitive teaching at these levels. The proponents of discovery learning based their proposal on Piaget's theory. This methodology had a great development in the 60s and 70s, and nowadays it focuses more on university teaching, emphasizing the active participation of students and, in the learning and application of science processes, it is postulated as an alternative to passive methods based on memorization and routine.
2. Inquiry: here students must explore for themselves; in a certain way, they must perform the role of the scientist or engineer in seeking the solution to a problem, discovering relationships and recording them for later treatment when solving a problem.
3. Connect: it is important to be able to identify the different connections between components, data, products, processes, etc. in such a way that they can be integrated into the solution of the problem.
4. Create: look for ideas, structure them in a plan, project or model that can be used to solve the initial problem.
5. Test: the ideas to verify the designed plan or model. It is important that the student can be sure of the correct solution to the practical problem that they set out to solve, and, for this, they must verify this proposal as many times as necessary.
6. Improve: think about how to improve the design, modify it and retest the ideas. The student must learn from mistakes and find ways to improve their solutions to practical problems.

Problem-Solving and Interactive Computational Simulations

Currently one of the most common methods for teaching science and technology at the university level is problem-solving (Dogru, 2008). A widely-used strategy with this approach is to organize didactic units articulated fundamentally as collections of problems. These problems must be carefully selected and sequenced in such a way that significant learning of the subject under study is achieved. Campanario and Moya (1999) point out that *'the word "problem" should be understood in a broad sense, since it includes, for example, small experiments, sets of observations, classification tasks, etc.'*. It should be noted that the internal dynamics of this strategy encourage self-regulated learning; in this way, when solving a problem is approached, an initial analysis is started so that the student develops his own mental model that describes the situation of the problem in question. In that first approximation, it is most likely that the mental model is not quite correct, is incomplete and has gaps. Also, it is likely that, in a second analysis, the student will discover new forms of solution, perhaps simpler or more complicated. In this solution process, it is important to analyze and determine the key factors to find the solution to the problem. When this process is carried out in a group and is well-organized in the classroom, it is extremely enriching for the learning of concepts by students, who, although they may be used to solving problems by applying mechanical algorithms without understanding laws and concepts, now, by analyzing and discussing in a group and in a collaborative way the possible steps to reach the correct solution, they will be able to build concepts and develop a useful resolution methodology. A distinctive feature of this methodology is that the complexity of the problems to be solved increases little by little and obviously increases the challenge and the time that must be invested to reach the solution. Throughout this process, the student is responsible for their own learning, which is why it is mentioned that the methodology is based on self-regulated learning.

The teacher must be careful to structure their collection of problems in such a way that certain specific concepts that are to be learned by the students are addressed and, little by little, should add other concepts until the students reach a maximum status of complexity where full knowledge of concepts and the management of resolution strategies are required. It is not intended that the student discovers concepts as is done in learning as research. *'The systematic use of problems would be intended to give relevance to such content, not to provoke its discovery'* (Campanario & Moya, 1999).

Among the advantages attributed to problem-based learning is that is more suitable than traditional transmission methods for the needs of students, since, in their professional practice, they will have to face more and more situations in the field of experimental sciences that provoke a search for solutions to problematic situations in real life. Given that this teaching strategy makes the application of theoretical knowledge to problem situations explicit, it clearly shows its importance and usefulness, and also contributes to an increase in intrinsic motivation for having found the solution to a problem. On the other hand, given that the student must make use of their knowledge learned in the classroom and that there is a continuous interrelation between theory and practical application, problem-based learning can achieve a better integration of declarative and procedural knowledge (Yeo & Gilbert, 2022).

One disadvantage of problem-based learning, and of the Interactive Lecture Demonstration, is the matter of time. The teacher must spend more time than normal to prepare the classes and in their execution, since the selection of problems must be carried out meticulously to include certain concepts of study, in addition to the fact that attractive problems, rich in context, whose sequencing arouses the interest and motivation of the students, must be selected. In the same way, a greater investment of time is also required from the student, and they are not always willing to spend more time solving problems. On the other hand, there is a variety of diverse approaches within the problem-based learning methodology, since the steps to follow when addressing problems vary from author to author. There is a proposal for solving problems with pencil and paper, problems in context, problems that require the use of ICT and computational resources but, despite everything, this is still a valid methodology. However, nowadays, the use of interactive computational simulations for teaching science is a very valuable resource for teachers (Cassam-Atchia & Rumjaun, 2022).

The Laboratory as a Teaching Strategy

The importance of the laboratory in educational research has been widely discussed by various authors from the end of the twentieth century (Hofstein & Lunetta, 2004). It is worth mentioning that, in the last twenty years, the interest of researchers in this has been growing, along with significant development of technology applied to education at all educational levels. So, at present, the laboratory plays a central role in science education and in technological education, due to the enormous potential that laboratories have to develop the research skills in students that are so necessary to address the problems of science and technology (Pokoo-Aikins et al., 2019).

The evolution of traditional laboratories into the 'extended laboratory' (Idoyaga et al., 2020) is one of the great advances that comes to restore the preponderant role that the laboratory has played in the teaching of science and that, nevertheless, little by little, was reduced in curricular times, in some cases due to financial constraints problems, in others due to trying to unify experimental disciplinary fields such as physics, chemistry and biology into one, with real laboratories even being replaced by virtual ones. The outbreak of the COVID-19 pandemic created scenarios of forced social distancing, leading to the closure of schools and opening cyberspace to Emergency Remote Teaching devices, thus reviving interest in remote, virtual and mobile laboratories. All this, coupled with simple experimental activities to perform at home, and computer simulations, led to the creation of the extended laboratory model, that is, the didactic and systemic use of devices and strategies to carry out experimental activities in digital educational environments, which seek to generate in students learning procedures, attitudes and concepts.

On the other hand, the idea that the laboratory provides students with opportunities to get involved in research and inquiry processes has endured; however, it is not clear how much the experiences of the students in the laboratory and their learning of scientific concepts influence, which has been one of the major problems of the last 40 years. During our time, the concept of the science laboratory has changed dramatically, as it has been greatly influenced by the use of technologies, as well as its association with inquiry learning. This is possible due to the role of the laboratory as a scene of scientific discoveries, and it is believed that, in a similar way to how the scientist discovers something of nature in the laboratory, so also the students find something that they have to discover when interacting in the laboratory. Now there are greater challenges to study and to see how they influence, for example, the role of the extended laboratory in the learning of science concepts. Similarly, the emergence of the STEM model that involves inquiry-based learning has strongly motivated the use of face-to-face and virtual science laboratories.

History and Philosophy of Science in Teaching

The contribution of science historians to teaching has played an important role in highlighting the social aspect of science beyond its technicalities. Likewise, there is the human factor in the construction of scientific knowledge, which can be motivating for students who are starting out in scientific study and who can understand that scientists were people like them, with the limitations of their time, and that, in a rudimentary way, they were able to make significant contributions to science by applying their abilities and investigative skills. Is it possible for a modern-day student to reproduce classic or transcendental science experiments to increase their learning? The answer is yes. It is strange to see how the writers of current high school textbooks generally ignore the original data of the key experiments in the history of science, in addition to including in their stories data, numbers and situations very different from those that occurred 300, 200 or 100 years ago. It is important that the teacher takes a historical approach with the students, so that the essence of such a process is not simply to expose the conclusions, but to show how they were reached and what options were discussed as being possible; thus, the historical process encourages thinking and, in many cases, experimentation.

The same happens with philosophy, since this, like history, is renewed and evolves with new applications, such is in the case of 'Philosophy for Children' applied to the teaching of physics (Mora, 2022). This methodology was created in the late 1960s by the American philosopher Matthew Lipman, with the aim of developing critical and creative thinking in elementary school children, to teach them to philosophize on current issues, through the development of philosophical skills. It should be noted that the development of the Philosophy for Children program was even expanded to teach adolescents, and today the teaching method is solidly structured and certified, from 3–18 years old. As this was a successful and proven method, over decades, for producing critical and creative thinking in children, the author sought to apply it to the teaching of science to pre-school children and very encouraging results were obtained.

Summary

The current dominant trend in science education and technology education is undoubtedly Active Learning, which uses reflection as a general method and active interaction between students under the mediation of the teacher. Reflection requires the identification of both central facts and open questions about the object of learning. The same goes for students' own ideas, emotions, resistances, values and preferences. In addition, common reflection in small groups helps students to learn about alternative perspectives. Discussion and debate help to reinforce correct concepts and discard incorrect ones. Among the most successful active methodologies, we have the Interactive Lecture Demonstration and the Project Method, which also has a strong application in STEM education, due to its integrating nature of scientific disciplines.

By using the STEM methodology, it is expected to achieve greater competitiveness and greater economic prosperity, in addition to being an index of a country's capacity to maintain sustained growth. Resorting to this educational model for teaching science and technology is not only to follow a fashion or passing trend, but to improve the economies of the countries. Among the challenges we have in STEM education are: (1) learning to work as a team; (2) that students learn to obtain relevant information and know how to handle it; (3) to encourage discussion and critical analysis; (4) to train students to carry out science and technology projects; (5) to train new teachers in STEM methodologies; and (6) to use ICT more to spread the STEM model around those regions of the world where economies are more precarious. One of the advantages of STEM education is that it promotes teamwork, leadership and communication with peers. Likewise, it helps students to solve real problems, learn from their mistakes, develop their creativity and logical thinking and the ability to improvise. We recommend that teachers look for the new advancements of Science Education research for having elements of effective instruction (Banilower et al., 2010).

In the case of learning through problems, it is one of the methodologies that abounds in university education and, given the nature of science and technology, there is a scenario of immediate application of concepts, ideas and procedures to solve the problems of everyday life. However, it is a methodology that requires a great investment of time by all the actors and, as it progresses in complexity, it produces greater motivation and acceptance in the students, as well as the development of critical and independent thinking skills that are not normally available and that are required in science education and technological education. In the line of teaching science through technology, we recommend explore and use the PhET interactive simulations (https://phet.colorado.edu/).

The laboratory is a very important science and technology learning resource, and is a valuable teaching tool to help develop research and inquiry skills in students, as well as promoting their involvement in scientific research. Furthermore, recent models of the school laboratory as a scientific research laboratory, coupled with the use of technologies, have led to the development of the extended laboratory model, where remote, virtual and mobile laboratories are involved, along with simple experimental activities at home. Again, computational simulations can be of great help, therefore we recommend the use of applets and interactive simulations like PhET project of Colorado University.

Finally, the use of the history of science and philosophy for teaching science and technology leads us to consider a new perspective of historical review in the classroom, which allows students to know the human side of scientists, as well as the social meaning of their discoveries. In addition to the motivating effect on the students, this has the encouraging effect of improving their learning by looking at how the original conclusions were reached. Similarly, the innovative application of the Philosophy for Children methodology to teach basic physics principles to kindergarten students was very encouraging.

Acknowledgement This chapter was written with the support of the research project SIP-20221757 *Active and Significant Learning of Physics through Physlets* of the Instituto Politécnico Nacional of Mexico.

References

Banilower, E., Cohen, K., Pasley, J., & Weiss, I. (2010). *Effective science instruction: What does research tell us?* (2nd ed.). RMC Research Corporation, Center on Instruction.

Callahan, B. E., & Dopico, E. (2016). Science teaching in science education. *Cultural Studies of Science Education, 11*, 411–418.

Campanario, J. M., & Moya, A. (1999). ¿Cómo enseñar ciencias? Principales tendencias y propuestas. *Enseñanza de las Ciencias, 17*(2), 179–192.

Cassam-Atchia, S. M., & Rumjaun, A. (2022). The real and virtual science laboratories. In B. Akpan, B. H. Zhang, B. Çavaş, T. J. Kennedy, & ICASE (Eds.), *Contemporary issues in science and technology education*. Springer Nature. In Press.

Dogru, M. (2008). The application of problem-solving method on science teacher trainees in the solution of the environmental problems. *Journal of Environmental & Science Education, 3*(1), 9–18.

Hofstein, A., & Lunetta, V. N. (2004). The laboratory in science education: Foundations for the twenty-first century. *Science Education, 88*(1), 28–54.

Idoyaga, I., Vergas-Badilla, L., Moya, C. N., Montero-Miranda, E., & Garro-Mora, A. L. (2020). El Laboratorio Remoto: una alternativa para extender la actividad experimental. *Campo Universitario, 1*(2), 4–26.

Kennedy, T. J., & Odell, M. R. L. (2022). STEM education as a meta-discipline. In B. Akpan, B. H. Zhang, B. Çavaş, T. J. Kennedy, & ICASE (Eds.), *Contemporary issues in science and technology education*. Springer Nature. In Press.

Li, Y., Wang, K., Xiao, Y., & Froyd, J. E. (2020). Research and trends in STEM education: A systematic review of journal publications. *International Journal of STEM Education, 7*(11), 1–16. https://doi.org/10.1186/s40594-020-00207-6

Meltzer, D. E., & Thornton, R. K. (2012). Resource letter ALIP–1: Active-learning instruction in physics. *American Journal of Physics, 80*(6), 478–496.

Mora, C. (2022). Early STEM implementation in PreK and kindergarten in Mexico. In S. D. Tunnicliffe & T. J. Kennedy (Eds.), *Play and STEM education in the early years international policies and practices* (pp. 363–379). Springer. https://doi.org/10.1007/978-3-030-99830-1

Pokoo-Aikins, G. A., Hunsu, N., & May, D. (2019). Development of a remote laboratory diffusion experiment module for an enhanced laboratory experience. In *IEEE Frontiers in Education Conference (FIE)* (pp. 1–5). IEEE. https://doi.org/10.1109/FIE43959.2019.9028460

Sokoloff, D. R., & Thornton, R. K. (1997). Using interactive lecture demonstrations to create an active learning environment. *The Physics Teacher, 35*, 340. https://doi.org/10.1109/1.2344715

Voelker, A. M., Thompson, T. E., & Vandeman, B. A. (2006). Research reviews in science education: An update. *Science Education, 64*(4), 569–578. https://doi.org/10.1002/sce.3730640407

Yannier, N., Hudsonkenneth, S. E., Koedingerkathy, R., Michnick, H.-P. R., Munakata, G., et al. (2021). Active learning: "Hands-on" meets "minds-on". *Science, 374*(6563), 26–30. https://doi.org/10.1126/science.abj9957

Yeo, J., & Gilbert, J. K. (2022). Producing scientific explanations in physics – A multimodal account. *Research in Science Education, 52*, 819–852. https://doi.org/10.1007/s11165-021-10039-1

Further Reading

Methods for teaching science. https://www.fizzicseducation.com.au/articles/methods-for-teaching-science/
50 best strategies for enhancing your science instruction. https://www.ber.org/seminars/course/CZZ/50-Best-Strategies-for-Enhancing-Your-SCIENCE-Instruction-Using-Cutting-Edge-Tech-Tools-and-Resources-Grades-6-12

Dr. César Mora is currently a Professor at the Research Center in Applied Science and Advanced Technology of the Instituto Politécnico Nacional (CICATA-IPN), Mexico. He was co-founder of the Latin American Physics Education Network (LAPEN) in 2005, and is associated member of the International Commission in Physics Education (ICPE) of the International Union of Pure and Applied Physics (IUPAP). He is a member of the National System of Researchers level I of Mexico. He is a physicist and philosopher and holds an MSc in Physics from CINVESTAV-IPN and a PhD in Physics from Universidad Autónoma Metropolitana Iztapalapa (2001).

Chapter 16
Pedagogical Content Knowledge in Science and Technology Education

Louise Lehane

Abstract This chapter looks at a construct in teacher knowledge known as pedagogical content knowledge (PCK), which has been viewed by many to be the 'missing paradigm' in teacher education research. The history of PCK is presented, recent conceptualisations of PCK are explored and another construct, known as technological pedagogical content knowledge (TPACK), is introduced. The recent COVID-19 pandemic has led to opportunities for both students and teachers to work through online platforms, therefore development of TPACK would be viewed as more important now than ever.

Teaching strategies that show well-developed knowledge of how to teach scientific content are explored and ways of capturing and measuring PCK and TPACK are presented. Throughout the chapter, the author will engage in reflective consideration for how PCK, in particular, has shaped her knowledge of teaching and, to that end, presents a new model to conceptualise PCK that includes consideration of current trends in science and technology education.

Keywords Pedagogical content knowledge · Technological pedagogical content knowledge · Teaching and learning · Content representation

Introduction

Pedagogical Content Knowledge was originally defined by Shulman (1986) as:

'For the most regularly taught topics in one's subject area, the most useful forms of representation of those ideas, the most powerful analogies, illustrations, examples, explanations, and demonstrations – in a word, the ways of representing and formulating the subject that make it comprehensible to others' (Shulman, 1986, p. 9).

L. Lehane (✉)
St Angela's College, Sligo, Ireland
e-mail: llehane@stangelas.ie

© The Author(s), under exclusive license to Springer Nature
Switzerland AG 2023
B. Akpan et al. (eds.), *Contemporary Issues in Science and Technology Education*, Contemporary Trends and Issues in Science Education 56,
https://doi.org/10.1007/978-3-031-24259-5_16

PCK signifies not only the amount of knowledge that a teacher has of the content, but also the organisation of that knowledge (Shulman, 1986). It is an amalgamation of knowledge of content, but also how to teach that content to make it understandable to others. A scientist, for example, would have very well-developed content knowledge, but may not necessarily have the knowledge to teach that content to others, therefore PCK is unique to the province of teachers (Shulman, 1986).

This chapter will critically examine PCK, from its inception to its development in research, and will provide opportunities to look at evidence-based teaching strategies, which, when used efficiently, show well-developed PCK. The concept of technological PCK will be introduced.

The chapter then presents an example of a tool used widely to capture PCK and has been adapted in this chapter to focus on technological pedagogical content knowledge (TPACK). Critical discussion and reflection on PCK, current trends and why it is so important to consider it in the planning and delivery of lessons are provided throughout the chapter.

Pedagogical Content Knowledge (PCK): An Historic and Current Theoretical Construct

The construct of PCK was originally presented by Lee Shulman as the 'missing paradigm' in educational research in the Presidential Address at the 1985 Annual Meeting of the American Educational Research Association. In his address, he proposed that there was an absence of focus on subject matter knowledge and an emphasis on teaching practices in the historical research. Essential questions in relation to knowledge were being avoided: questions like *'how do teachers decide what to teach, how to represent it, how to question students about it and how to deal with problems of misunderstanding?'* (Shulman, 1986, p. 8). Furthermore, the knowledge components of subject matter knowledge and pedagogical knowledge were often considered in isolation from each other and both needed to be viewed as mutually inclusive in order to allow for the transformation of effective teaching and learning in the classroom – in other words, to make the material that you are teaching understandable to others (Shulman, 1986).

Since its original inception, it has informed the direction of significant research in education and has undergone transformations in terms of its reconceptualisation by many distinguished scholars involved in PCK research. In order to understand its evolution, it is crucial to present the components of PCK as envisaged by various scholars dedicated to the field of PCK research. The following (Table 16.1) presented by Lee and Luft (2008) provides a summary of such components, as scholars seek to find a conceptualisation of PCK that best provides for Shulman's original vision of what PCK is.

These are all very much historic conceptualisations of PCK. The model by Magnusson et al. (1999) has been used extensively in research and an adapted version will be presented at the end of this chapter, with consideration for technological applications.

Table 16.1 Historic conceptualisations of PCK adapted from Lee and Luft (2008, p. 1346)

Knowledge of	Subject matter	Representations and instructional strategies	Student learning and conceptions	General pedagogy	Curriculum and media	Context	Purpose	Assessment
Shulman (1987)	a	PCK	PCK	a	a	a	a	b
Tamir (1988)	a	PCK	PCK	a	PCK	b	b	PCK
Grossman (1990)	a	PCK	PCK	a	PCK	a	PCK	b
Marks (1990)	PCK	PCK	PCK	b	PCK	b	b	b
Cochran et al. (1993)	PCKg	b	PCKg	PCKg	b	PCKg	b	b
Fernandez-Balboa and Stiehl (1995)	PCK	PCK	PCK	b	b	PCK	PCK	b
Magnusson et al. (1999)	a	PCK	PCK	a	PCK	a	PCK	PCK
Carlsen (1999)	a	PCK	PCK	a	PCK	a	PCK	b
Loughran et al. (2001)	b	PCK	PCK	b	PCK	b	PCK	PCK

a distinct category in the knowledge base for teaching, *b* not discussed explicitly, *PCK* Pedagogical Content Knowledge, *PCKg* Pedagogical Content Knowing

There have been many recent expansions and interpretations of the model of PCK, so much so that Barrett and Green (2009) state that there are as many variations of the term PCK as there are researchers interested in it. Indeed Loughran et al. (2006) consider that some examples of PCK bear little resemblance to the construct originally developed by Shulman (1986). While the above table provides historical conceptualisations that have been presented in the literature on PCK since its inception into the research realm, Table 16.2 below developed by Lehane (2016) provides

Table 16.2 Recent conceptualisations of PCK (Lehane, 2016)

Literature source	PCK components
Ball, Thames and Phelps	Knowledge of: (a) subject area, for example being able to write up a report for a laboratory experiment; (b) content and students, which refers to knowing the students, for example their commonly held misconceptions as well as knowing the subject matter; and (c) content and teaching.
Henze, van Driel and Verloop	Knowledge of: (a) instructional strategies; (b) knowledge about students' understanding; (c) knowledge about ways to assess students' understanding; and (d) knowledge about goals and objectives of the topic in the curriculum.
Park and Oliver	Orientations towards science teaching. Knowledge of: (a) students' understanding of science; (b) science curriculum; (c) instructional strategies and representations; and (d) assessment of science learning. Teacher efficacy. Model reflects interactivity and coherence between components. This model is referred to as the ***hexagon model***.
Hagevik, Veal, Brownstein, Allan, Ezrailson and Sean	Knowledge of: (a) context, curriculum and assessment; (b) instructional strategies and representations of teaching science; (c) student learning; and (d) knowledge of student understanding about science concepts.
Mavhunga	Knowledge of: (a) students' prior knowledge including misconceptions; (b) curricular saliency; (c) what makes a topic easy or difficult to understand; (d) representations including analogies; and (e) conceptual teaching strategies.
Types of PCK	
Veal and McKinster	General PCK, Domain-specific PCK and Topic-specific PCK.
Lee and Luft (2008) – drew on the work of Gess-Newsome	Transformative (synthesis of all the knowledge required to be an effective teacher) versus Integrative PCK (the knowledge domains of subject matter, pedagogy and context exist as separate entities).
Daehler and Heller	Espoused (teacher knowledge) and enacted PCK (what happens in the classroom). A teacher's espoused PCK does not necessarily mean that it will be enacted in the classroom (Aydeniz & Kirbulut, 2011). Park, Jang, Chen and Jung (2011) considered two similar dimensions of PCK: understanding (what a teacher knows) and enactment (what a teacher does in the classroom).

a summary of some of the more recent conceptualisations and types of PCK found within the relevant literature:

The varying conceptualisations of PCK presented in both tables highlight the complexities around defining and understanding what teacher knowledge is and on what the focus should be.

In recent years, the concept of technological pedagogical content knowledge (TPACK) has been developed and utilised in research and practice and will be discussed in more detail later on in the chapter. It must be mentioned that, while science and technology education are the focus of this chapter, PCK as a construct can be considered in the teaching of all subjects.

Pedagogical Content Knowledge: Why Is It Important?

In Shulman's address, he discussed the negative association with teaching and referenced George Bernard Shaw's infamous aphorism that 'He who can, does. He who cannot, teaches' (Shaw, 1903, cited in Shulman, 1986). Shulman's research led him to decipher a distinction between content knowledge and pedagogical method in the hope that the findings of his research would reverse the current negative associations with teaching, so that it could be viewed as the complex activity that it is (Shulman, 1986).

In order to be able to distinguish the knowledge of, say, a scientist from that of a science teacher, it is important to consider what enhanced knowledge a science teacher may have that a scientist does not necessarily possess. The ability to be able to provide understandable explanations to specific students, to be able to address diverse needs in the classroom and to provide opportunities for specific pedagogies that enhance the learning of students, is an example of how a scientist's knowledge may vary from that of a science teacher, effectively their knowledge of science content and how to teach that content to make it understandable to others, and that is PCK. Such knowledge is fundamental to the students' learning experience and that is why it is so important to consider it in both initial teacher education and for in-service teachers out on practice.

From a pre-service teacher's perspective, this author has worked in initial teacher education for 12 years, using PCK as the central tenet in her teaching. She continually tries to emphasise the equal importance of understanding the content that one is to teach, but how to teach it in such a way that it is made comprehensible to others – this is the essence of PCK. Despite her attempts, pre-service teachers struggle to focus on that amalgamation of different knowledge domains and therefore it is necessary to find concrete ways of making PCK part of a pre-service teacher's consideration for how they plan to teach. Later in this chapter, a tool to capture PCK will be examined.

PCK is, of course, crucial for in-service teachers; however, with the limited classroom experience that pre-service teachers have on the 'other side of the classroom desk', and their often-tenuous journey transitioning from a student to a teacher, having a framework to guide their developing knowledge of how to teach is warranted.

Introducing Technological Pedagogical Content Knowledge (TPCK/TPACK)

Technological pedagogical content knowledge (TPCK), which is now referred to TPACK (technology, pedagogy **and** content knowledge), is a more recent concept that is effectively an extended conceptualisation of PCK to include technology knowledge (Harris et al., 2009). The three bodies of knowledge of content, pedagogy and technology knowledge, and how they interact with each other, produces a flexible approach to teaching that allows for the purposeful integration of technology into a teacher's repertoire.

It is considered crucial for effective teaching with technology, both in science and technology education, as interaction with technology can promote critical thinking and other key skills synonymous with both science and technology as school subjects. The recent COVID-19 pandemic and the subsequent switch to online teaching has illustrated the need to include technological applications in our pedagogical approaches, regardless of whether the setting is within a school context or not. However, the swift nature of having to adapt to online teaching has led to teachers' TPACK being tested, with varied impact on student learning. Significant research has already taken place on the impact of online teaching on student learning, and perspectives of both teachers and students show mainly negative associations with online teaching (Nambiar, 2020). The key question is, why can both students and teachers hold negative orientations towards online teaching? It can be suggested that the pedagogical approaches employed by the teachers and the low levels of self-efficacy with using online platforms can affect the experiences of both teachers and students, both of which are intimately linked to TPACK.

There are also additional challenges in teaching with technology, as identified by Harris et al. (2009). Social and contextual factors, such as poor infrastructure around technology available to students both at home and within the classroom, would be seen as particular challenges, and something that the use of online teaching due to school closures during the COVID-19 pandemic identified was the social gap of technology accessible to different students. What is only now becoming apparent is the social divide and, as a result, the learning divide between students with and without appropriate access.

An additional challenge presented by Harris et al. (2009) is the experience of teachers of using technology, which is in effect their TPACK. TPACK, like PCK, develops with experience and reflection on experience so, if teachers are to develop their TPACK, they need to use professional learning opportunities to engage in reflection.

Examining How PCK and TPACK Relate to Science and Technology Teaching

Significant research has looked at ways of conceptualising PCK, and measuring and capturing PCK, with more recent research looking at particular aspects of PCK. For example, Lehane (2019) examined how a tool used to capture PCK, known as the content representation (CoRe) tool, could capture pre-service teachers' understanding of nature of science. Other research has focused on the teaching of particular topics (e.g., Gencer and Akkus (2021), who focused on the interactions between chemical species and states of matter through a PCK lens), or in the teaching of particular scientific process skills (e.g., Lehane, 2016).

Other research has looked at the idea of enacted PCK versus espoused PCK, in which the former looks at PCK in action in the classroom, while the latter examines teachers' perceived PCK, which may not necessarily transfer into classroom practice (Lehane, 2016).

More recent studies have begun to use the CoRe tool to investigate early childhood teachers' collective PCK and personal PCK (Buldu & Buldu, 2021). The CoRe tool is often used in group settings where teachers collectively present their ideas of how they would teach particular topics, and the very nature of this collaborative effort and sharing of ideas can enhance their own PCK construction (Lehane, 2016).

Ways of measuring PCK have been a key focus of research over the years, with new instruments being developed and validated (see He et al., 2021). The overarching rationale for finding ways to capture and measure PCK is that it is an elusive construct and, in order for a teacher's PCK to result in impact in the classroom, it is crucial to find ways of making it visible.

The research into teachers' PCK is tending to focus more recently on pre-service teachers, rather than in-service teachers, perhaps indicating the need to view this as a necessary framework to develop their understanding of the key knowledge components needed to be an effective classroom practitioner.

Teaching Strategies that Suggest Well-Developed PCK and TPACK

The use of evidence-based teaching strategies in the classroom would suggest high levels of PCK and TPACK. The following section presents and describes some strategies that can be used that can have a technology focus in their implementation. All of these strategies would be seen as having high effect sizes according to Petty (2009), which show evidence of enhanced achievement levels of learners engaged in such strategies, compared to other learners.

- *Jigsaw methodology*
- The jigsaw methodology is a co-operative learning activity where students work in 'expert groups' to complete a task assigned to them, often engaging in a problem-solving approach (see Chap. 18). They then return to their 'home groups' and share their learning from the information garnered from engaging in the task in their respective expert group. Each member of a 'home group' has come from their own 'expert group', where they have completed their own individual task to provide the other members of the 'home group' with key information from same. The key benefit of the jigsaw methodology is that students are constructing their own knowledge while working in groups, learning key skills such as communication and working with others. Additionally, it provides students with a sense of responsibility that they bring back accurate information to their 'home group' members. Finally, due to the fact that each task results in different information being generated and summarized, the jigsaw methodology can be used to teach a significant amount of content. From a technology perspective, breakout rooms on learning platforms can be used to assist with this. Also, tasks could involve online research for specific tasks.
- *Interactive video methods*
- A key technology-based methodology would be the use of interactive videos, which can be used in tangent with other teaching strategies such as note-taking and summarizing, both described in due course.
- *Concept mapping*
- Concept maps are graphical organizers, which provide a way of representing students' knowledge. The content related to a particular topic is presented in a hierarchical structure, from general to more specific concepts (both presented in nodes) related to a topic, with linking phrases, cross-links and propositions between the concepts. A concept map could be used to summarize information garnered from online research conducted by the students. Like all graphical organizers, concept maps summarize and synthesize key concepts related to a topic, but the presence of linking phrases, cross-phrases, etc., where one has to make a connection between one concept and another, requires deeper thinking and a well-developed knowledge of the topic.
- As well as being used at the end of a research task, they could also be potentially used as a study tool or an assessment tool, for example.
- *Note-taking and summarizing*
- Note-taking is a crucial skill for students to learn, but often it is approached in a didactic way by students taking down notes that the teacher has provided, without opportunity for students to think about what they are writing down. Changing this approach slightly by having students making notes in their own words allows for them to process the information learned in their own way. From a technological perspective, students can engage in online research and, from this, create

their own summary notes. Additionally, online platforms could then be used to share these notes, allowing the teacher to provide appropriate feedback.

- *Reciprocal teaching*
- This is a strategy used to develop reading comprehension skills and follows a particular cycle during a reading task. It includes five stages: predicting, silently reading, questioning, clarifying and summarizing.
- First, the classroom teacher predicts the content of a paragraph within a piece of text; they then get all students to read a piece of text silently. The teacher then questions the students on particular content in the text, which is followed by the teacher clarifying any misconceptions that the students may have. Next, the teacher summarizes the paragraph in a short phrase or sentence. After this, a student acts as the 'teacher' and the whole cycle starts again with the next paragraph, where they first predict what they think the focus of the next paragraph will be.
- This is an excellent approach to use both in the physical and online classroom environment. Additionally, if used online, the piece of text can be shared on screen and key points highlighted to help guide the readers as they work through the text.
- *Decisions, decisions*
- This teaching strategy is a series of learning games that are sometimes called 'manipulatives'. Students are given a set of cards containing words, visuals, numbers, etc.; they are then asked to sort, sequence, match, group and classify. From a virtual perspective, students could drag and drop text boxes and diagrams, etc. The management of this activity can vary, with students either completing these tasks individually or in pairs. Online platforms provide breakout rooms where respective students can work together and then come back to the main room to share their findings.
- *Flipped classrooms*
- The flipped classroom approach is widely used internationally. It consists of students engaging in specific homework tasks, for example, getting students to research information on a particular concept, e.g., to research the effect of pH and temperature on the rate of enzyme activity, which is commonly found on biology syllabi internationally. Students then, in class, present their findings from looking at secondary data available online. With the flipped classroom approach, the majority of work is done by the student independent of the classroom environment, which is subsequently used to share what they have learned. The flipped classroom, from a psychology of learning perspective, also has the benefit of providing students with autonomous learning opportunities where they construct their own knowledge and, therefore, it enhances their understanding of a particular idea according to relevant research in the area.

Examining the Place of PCK and TPACK in Initial Teacher Education

This section will explore how PCK awareness can be used to foster the professional development of pre-service teachers in initial teacher education.

A previous section has referred to a PCK tool developed by Loughran et al. (2006) to capture PCK. A CoRe is completed for individual topics. It contains a number of pedagogical prompts on the left-hand side, and consideration of all of these in a teacher's planning and delivery of a topic can significantly enhance the students' learning experience. The person or persons completing the CoRe need to firstly identify what they believe are the 'Big Ideas' in a particular topic, and the pedagogical prompts unpack the Big Ideas. Big Ideas refer to the science ideas that teachers view as crucial for students to develop their own understanding of a particular topic. An example of a Big Idea from the topic of chemical reactions would be: *'A chemical reaction is when 2 or more substances come together and have an effect on one another to produce different products'* (Lehane, 2016).

The CoRe tool has been adapted in several studies for different research purposes; for example, Lehane (2016) adapted the CoRe tool to have a scientific enquiry focus. To that end, this author would suggest that the CoRe could be adapted to focus on developing TPACK and is presented in Fig. 16.1 below.

Other research has focused on measuring PCK through tools such as surveys and tests. However, in terms of effectiveness, it can be argued that capturing PCK would have a more significant impact on teachers' professional development and in turn student learning. The reason for this is that tools such as CoRe make visible teachers' knowledge of teaching particular topics. It captures all aspects both of a teacher's pedagogical and content knowledge. Additionally, the CoRe can be completed by teachers in groups, thereby allowing CoRe construction to be a professional learning opportunity as teachers listen to each other's contributions (Lehane, 2016).

	Big Idea	Big Idea	Big Idea	Big Idea
What do you intend students to learn about this idea?				
Why is it important for students to know this – consider specific relevance to everyday life				
What else do you know about this idea (that you do not intend students...)?				
Difficulties/limitations connected with teaching this idea				
Knowledge about students' thinking that influences your teaching of this idea. Consider students' understanding of information technology in your response				
Other factors that influence your teaching of this idea				
Teaching procedures with specific ICT focus				
Specific ways of ascertaining students' understanding or confusion around this idea (include likely range of responses)				

Fig. 16.1 Adapted CoRe

Author's Reflection on Working in PCK Research

I have work in PCK research with pre-service teachers for 12 years. My work has mainly looked at using the PCK tool, CoRe (described earlier on in the chapter), as a lens to capture their PCK, but also to allow the pre-service teachers to socially construct and develop their PCK through working within a group. Pre-service teachers involved in their CoRe sessions have identified significant benefits from their involvement. These benefits include viewing a CoRe as a lesson planning tool to assist them in thinking about how they would represent material to make it comprehensible to others (Lehane, 2016). Pre-service teachers also identified it as a means to document their progress as they develop their own teacher identity, and as a way to think critically by working together as opposed to being given the information by their teacher educators (Lehane, 2016).

Interestingly, I have worked in two universities since becoming a teacher educator and what remains the dominant concern for pre-service teachers is the teaching practicum experience. Despite this concern, those involved in my studies have vocalised that, because of their enhanced understanding of PCK, they can now make the informed connection between what to teach and how to teach it.

There are however some challenges associated with having PCK as the guiding framework for teacher training. I would argue that a teacher's PCK and TPACK need to develop organically; teachers need to see the value in understanding the importance of PCK in enhancing student learning, thereby teacher attitude and motivation to develop can be seen as a significant challenge. Teachers need to be aware of what their own PCK looks like and the CoRe tool can make visible their knowledge of the content and how to teach that content. It can however be a discomforting experience to reflect on your own knowledge as a teacher, but it is a necessary practice in order to enhance the learning of students. We talk about self-assessment as being a crucial part of formative assessment for students, yet we do not seem to routinely self-assess our own knowledge. I would argue that this should become a more routine practice in our own professional development.

Summary Thoughts

Since its inception, PCK has been a key focus of both research and curriculum policy, but the question is, where can we go now with PCK and, indeed, TPACK? John Settlage, in 2013, wrote an article entitled *On Acknowledging PCK's Shortcomings* and provided some interesting perspectives on how PCK is a *'persistent but unfulfilling notion'*, writing that it sparkles but offers little substance (Settlage, 2013, p. 2). Rarely has the literature critically examined PCK and offered negative perspectives. He does concede, however, that when one focuses on student learning as opposed to teacher learning with respect to PCK, this does have some merit (Settlage, 2013).

Reading the work of Settlage has provided me with a reflexive positioning on whether or not I truly believe that PCK needs to be at the heart of pre-service teacher education. Does PCK have a future in our practice? I would argue yes, but I agree with Settlage in terms of its need to be at the fore of documents specifying exemplar tools and practices for science teachers (Settlage, 2013), and that is where tools such as the CoRe tool can be used to draw out PCK and ultimately do what Settlage considers is missing from PCK research – the focus on student learning.

It is also necessary to discuss the model of an initial teacher education programme. For example, when I was training over a four-year concurrent training programme, I was taught pedagogy and content separately and, as a result, I did not see the importance of considering this amalgam of knowing the content and how to teach it. I was often learning the scientific content with students from other courses, therefore, there was no opportunity for discussion of how particular content could be taught in the classroom, i.e., how a teacher could make the material comprehensible to others.

If it had been explained to me while training, I believe that I would have seen the value in such considerations. That is perhaps something to consider going forward, breaking down the wall of theory and practice and allowing pre-service teachers to recognise the importance of their own knowledge development, without the 'academic tagline' that can sometimes be a barrier to their learning.

Furthermore, PCK as a construct is crucial for practicing teachers, particularly with an ever-changing understanding of how students learn. For example, inclusive education is viewed as being the gold standard of a teacher's planning and delivery in the classroom, but is an evolving framework. Therefore, PCK awareness needs to diversify to consider the current trends in education and I think that, by re-conceptualising PCK, this can be achieved.

To that end, I have provided a re-conceptualised model of PCK for consideration, which is an extension of the model developed by Magnusson et al. (1999) and is presented in Fig. 16.2 on the next page.

I have presented the above figure in a cylindrical model, as I feel this represents the relationship between all of the components of PCK and how one component informs another. A teacher may have very good knowledge of student-led, evidence-based teaching strategies, but this would need to be informed by their knowledge of students' understanding of science, for example.

Knowledge of context is an important consideration here, as a teacher's PCK can vary with different class groups, class settings and the challenges and opportunities that some groups or settings present. PCK develops with experience, but there is a need to consider both reflection *of* action and reflection *in* action in promoting PCK development. Reflection of action is looking back after an experience, while reflection in action is about looking at one's practice during the experience. Both are crucial for PCK development.

It is hoped that this chapter has provided the reader with some awareness of the importance of considering PCK in teachers' practice, both planning and delivery. It is important to think about your experiences as a student: who did you perceive to be the 'good' teacher? Was it the teacher who was patient or kind, or who

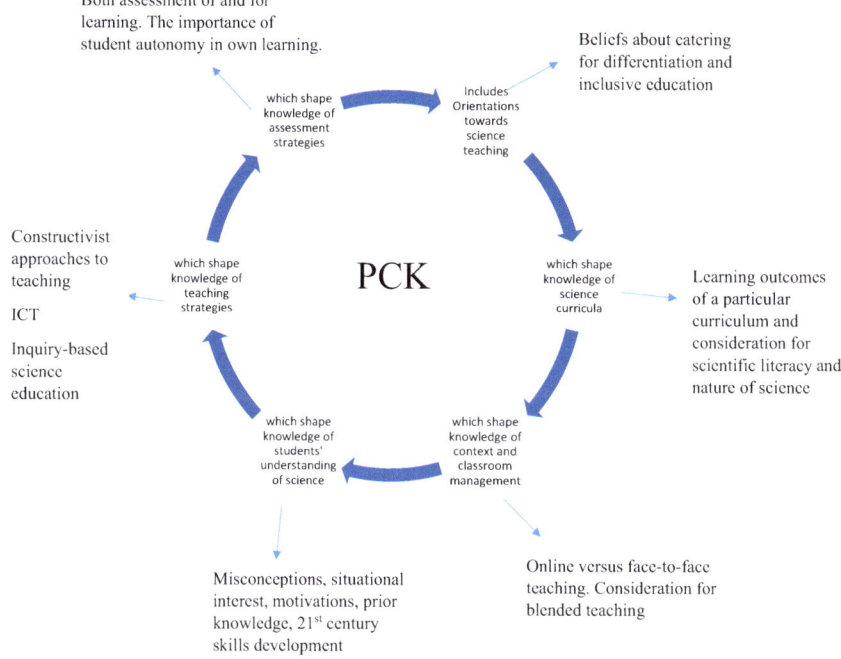

Fig. 16.2 PCK model. (Adapted from Magnusson et al. (1999))

demonstrated knowledge on the content, or who controlled the class well, or who made science fun? Was it a combination of some or all of these, perhaps? That is in essence what PCK is but, crucially, a good teacher does not just come to class and teach; an effective teacher recognises the complex nature of the learning experience and I truly believe that using the CoRe tool can help teachers plan appropriately and where, in turn, each learner can learn to the best of their ability.

Summary

In this chapter, I have discussed the meaning and origins of PCK as an academic construct. I have also focused on the different types of PCK and the importance of understanding PCK and how it can be practically considered in the classroom context. I then moved on to technological PCK (TPACK) and used presented an adapted PCK framework to consider TPACK. The discussion moved on to how both TPACK and PCK can be considered in science and technology teaching as well as in initial teacher education. Finally, the chapter looked at PCK research and the author's experience of working in same and where the research can now evolve and how PCK conceptualisations can be adapted to consider technology within same, through the presentation of an adapted model of PCK.

Recommended Resources

Darling-Hammond, L. (2008). Teaching and learning for understanding. In L. Darling-Hammond, B. Barron, P. D. Pearson, A. H. Schoenfield, E. K. Stage, T. D. Zimmerman, G. K. Cervetti, & J. L. Tilson (Eds.), *Powerful learning: what we know about teaching for understanding*. Jossey-Bass.

Hattie, J. (2009). *Visible learning: A synthesis of over 800 meta-analyses relating to achievement*. Routledge.

Loughran, J., Berry, A., & Mulhall, P. (2006). *Understanding and developing science teachers pedagogical content knowledge*. Sense Publications.

Petty, G. (2009). *Evidence based teaching: A practical approach* (2nd ed.). Nelson Thornes Ltd.

References

Barrett, D., & Green, K. (2009). Pedagogical content knowledge as a foundation for an interdisciplinary graduate program. *Science Educator, 18*(1), 17–28.

Buldu, E., & Buldu, M. (2021). Investigating pre-service early childhood teachers' cPCK and pPCK on the knowledge used in scientific process through CoRe. *SAGE Open, 11*(2), 1–16.

Carlsen, W. S. (1999). Domains of teacher knowledge. In J. Gess-Newsome & N. G. Lederman (Eds.), *Examining pedagogical content knowledge: PCK and science education* (pp. 113–144). Kluwer.

Cochran, K. F., DeRuiter, J. A., & King, R. A. (1993). Pedagogical content knowing: An integrative model for teacher preparation. *Journal of Teacher Education, 44*(4), 263–272.

Fernandez-Balboa, J., & Stiehl, J. (1995). The generic nature of pedagogical content knowledge among college professors. *Teaching & Teacher Education, 11*(3), 293–306.

Gencer, S., & Akkus, H. (2021). The topic-specific nature of experienced chemistry teachers' pedagogical content knowledge in the topics of interactions between chemical species and states of matter. *Chemistry Education Research and Practice, 2*, 498–512.

Grossman, P. L. (1990). *The making of a teacher: Teacher knowledge & teacher education*. Teachers College Press.

Harris, J., Koehler, M. J., & Mishra, P. (2009). What is technological pedagogical content knowledge? *Contemporary Issues in Technology and Teacher Education, 9*(1), 60–70.

He, P., Zheng, C., & Li, T. (2021). Development and validation of an instrument for measuring Chinese chemistry teachers' perceptions of pedagogical content knowledge for teaching chemistry core competencies. *Chemistry Education Research and Practice, 22*(2), 513–531.

Lee, E., & Luft, J. A. (2008). Experienced secondary science teachers' representation of pedagogical content knowledge. *International Journal of Science Education, 30*(10), 1343–1363.

Lehane, L. (2016). *Exploring the development of Irish pre-service science teachers' scientific inquiry orientations using a pedagogical content knowledge lens within a targeted learning community*. Unpublished Thesis (Ph.D.), University of Limerick.

Lehane, L. (2019). *Using a PCK lens to capture pre-service science teachers' internalised knowledge of nature of science*. Paper presented at the ASERA Annual Meeting, New Zealand, July 2019.

Loughran, J., Milroy, P., Berry, A., Gunstone, R., & Mulhall, P. (2001). Documenting science teachers' pedagogical content knowledge through PaP-eRs. *Research in Science Education, 31*, 289–307.

Loughran, J., Berry, A., & Mulhall, P. (2006). *Understanding and developing science teachers' pedagogical content knowledge*. Sense Publications.

Magnusson, S., Krajcik, J., & Borko, H. (1999). Nature, sources, and development of pedagogical content knowledge for science teaching. In J. Gess-Newsome & N. G. Lederman (Eds.), *Examining pedagogical content knowledge: The construct and its implications for science teaching* (pp. 95–132). Kluwer.

Marks, R. (1990). Pedagogical content knowledge: From a mathematical case to a modified conception. *Journal of Teacher Education, 41*(3), 3–11.

Nambiar, D. (2020). The impact of online learning during Covid-19: Students' and teachers' perspectives. *International Journal of Indian Psychology, 8*(2), 783–793.

Petty, G. (2009). *Evidence based teaching: A practical approach* (2nd ed.). Nelson Thornes Ltd..

Settlage, J. (2013). On acknowledging PCK's shortcomings. *Journal of Science Teacher Education, 24*, 1–12.

Shulman, L. (1986). Those who understand: Knowledge growth of teachers. *Educational Researcher, 15*(2), 4–14.

Shulman, L. S. (1987). Knowledge and teaching: Foundations of the new reform. *Harvard Educational Review, 57*, 1–22.

Tamir, P. (1988). Subject matter and related pedagogical knowledge in teacher education. *Teaching and Teacher Education, 4*(2), 99–110.

Dr. Louise Lehane is a Lecturer in Education at St Angela's College, Sligo, Ireland which is currently pending formal incorporation into the Atlantic Technological University. She lectures in the areas of General and Science Pedagogics and Sociology of Education.

She is a qualified science teacher and, following the completion of her initial teacher education programme, embarked on a PhD. Her thesis was focused on the use of a pedagogical content knowledge (PCK) lens to capture pre-service science teachers' scientific enquiry orientations within a professional learning community. Her main research interests include PCK, scientific enquiry, nature of science, curricular policy and the different stages of professional development.

Chapter 17
Stimulating Students' Mechanistic Reasoning in Science and Technology Education Through Emerging Technologies

Vickren Narrainsawmy and Fawzia Narod

Abstract This chapter focuses on how emerging technologies have the potential to stimulate students' mechanistic reasoning in science and technology. The work provides insight into the constituents of mechanistic reasoning and the accompanying challenges encountered by students to develop this important dimension of scientific and technological thinking and reasoning. As abstraction is ubiquitous in science and technology, concepts with varying degrees of abstraction (simple to complex and concrete to abstract) are used to explain phenomena. Developing mechanistic reasoning is challenging because students are required to progress through an increased level of sophistication in reasoning that entails using concepts with an increasing degree of abstraction when describing mechanisms for phenomena. Extended reality (XR) technologies, an umbrella term for augmented reality (AR), mixed reality (MR) and virtual reality (VR) among other emerging technologies, have the potential to stimulate students' mechanistic reasoning due to their ability to translate abstract objects and relations, typically represented in textual forms, into animated representations in a virtual environment. The three 'realities' differ on the reality-virtuality continuum as well as on the immersion and interaction spectra. Students' level of psychological immersion and interaction is maximum with VR; thus, VR has the full potential to stimulate students' mechanistic reasoning.

Keywords Mechanistic reasoning · Emerging technologies · Abstraction · Extended reality · Augmented reality · Mixed reality · Virtual reality

V. Narrainsawmy · F. Narod (✉)
Department of Science Education, Mauritius Institute of Education, Moka, Mauritius
e-mail: vickren.narrainsawmy@mie.ac.mu; f.narod@mie.ac.mu

© The Author(s), under exclusive license to Springer Nature
Switzerland AG 2023
B. Akpan et al. (eds.), *Contemporary Issues in Science and Technology Education*, Contemporary Trends and Issues in Science Education 56,
https://doi.org/10.1007/978-3-031-24259-5_17

237

Introduction

Science and technology are an integral part of a country's development, as scientific excellence and technological innovation are central to the performance of a country's economic growth and the improving conditions of its citizens. Science and technology, though interdependent, are highly interconnected due to the scientific nature of technology and the technological aspects of science. For a further conceptual understanding of the nature of science and the nature of technology, please refer to Chap. 2. From an educative perspective, the interconnectedness between science and technology enables the natural integration of the two disciplinary domains into one: science and technology education. Science and technology education empowers students with the required competencies to prosper in schools and beyond. The broad objectives of contemporary science and technology education are to prepare students for scientific and technological professions which will be of higher demand, for future career opportunities in a wide range of other areas that will require scientific and technological skills, and for an interest in, to appreciate its value in society. Please see Chap. 5 on curriculum design in science and technology education at international level for an in-depth description of the purposeful, deliberate, and systematic organization of science and technology curriculum. Science and technology education is discerned as crucial to achieving the desired future workforce competencies and future educated citizens.

An existing gap in science and technology education is between the ongoing traditional teaching practices and the development of future workforce competencies. With the unprecedented challenges of the twenty-first century and the prospective opportunities, increasing attention is generated among the education research community on how educators can empower students with the ability to think and reason scientifically and technologically, thereby better equipping the students with the twenty-first century workforce competencies. To bridge the existing gap, studies have reported that educators need to engage students in knowledge-based reasoning classroom practices for the students to effectively construct explanations of natural and technological phenomena from prior learning experiences. An important dimension of scientific and technological practice is reasoning about a mechanism (or mechanistic reasoning). Mechanistic reasoning is an important dimension of scientific practice and a central dimension of science curricula (De Andrade et al., 2022).

Mechanistic reasoning is critical to disciplined inquiry in the science and technology domain, as it accounts for the underlying cause-effect relationship of a phenomenon. Mechanistic reasoning enables students to construct meaning from the underlying factors and relationships that give rise to the phenomenon, thus explaining how and why this particular phenomenon occurs. Moreira et al. (2019) and Caspari and Graulich (2019) have stated that students face difficulties in developing such a type of reasoning. Developing students' mechanistic reasoning during the teaching and learning process is undeniably a challenge for educators. Thus, there is a need to reconsider how mechanistic reasoning could be developed among students in an age of technological advancement. Educators can leverage emerging technologies as innovative and creative pedagogical resources to develop students' mechanistic reasoning. For further reading on creativity and innovation in science and technology education, please refer to Chap. 19.

In view of the above discussion, this chapter has been conceptualized to provide a discourse on how students' mechanistic reasoning can be stimulated with the application of emerging technologies as educational resources. The approach to the discourse is 'domain-oriented', as it uses information from the philosophy of science and technology as a framework to examine how specific characteristic features of emerging educational technologies can potentially stimulate students' mechanistic reasoning. The rationale for this approach to the discourse is that many studies have been conducted on the opportunities and challenges of emerging technological approaches to science and technology education, but have ignored the complex conditions under which learning occurs. Thus, we use information from the philosophy of science and technology for shedding light on how specific characteristic features of emerging technologies have the potential to stimulate students' mechanistic reasoning. An analysis of what constitutes a mechanism and reasoning about a mechanism from the philosophy of science and technology is used as a lens to examine the specific characteristic features of emerging technologies.

The discourse is, therefore, guided by the following questions:

- What constitutes mechanistic reasoning in science and technology education?
- How can emerging technologies stimulate students' mechanistic reasoning in science and technology education?

Answering the questions is highly relevant for theory and practice, as we are almost midway through the time for the Sustainable Development Goals (SDGs) set by UNESCO to be achieved (by 2030). For a comprehensive understanding of sustainable development goals and science and technology education, please refer to Chap. 10. Also, the world has been severely impacted by an unprecedented COVID-19 pandemic situation, which has shown more than ever how the future is unpredictable. With the rapid advances of technology, the advent of Industry 4.0 and the twenty-first century VUCA (Volatile, Uncertain, Complex and Ambiguous) world, learners of today cannot be prepared for tailor-made jobs, as no one is certain about the jobs that will be created in the future. More than ever, education now needs to endow twenty-first century learners with the necessary competencies that will prepare them for flexibility and lifelong learning, and to be ever-ready to adapt and perform effectively in new jobs and new work environments. On the other hand, several benefits have been attributed to mechanistic reasoning, namely that it '*contributes significantly to understanding both the designed world and the natural world*' (Bolger et al., 2009) and that '*Exposure to mechanistic reasoning, even during the preschool period, may be useful to facilitate learning and advance children's cognitive reasoning abilities*' (Kurkul et al., 2021). Given the fact that digital transformation and Industry 4.0 are expected to drastically impact the nature of future jobs, it would be helpful to understand how emerging technologies can promote the development of mechanistic reasoning amongst learners in science and technology education. This would enlighten educators, teacher educators and other relevant stakeholders on how emerging technologies can best prepare learners to operate successfully as twenty-first century citizens and to work in the unknown and unpredictable future job market.

What Constitutes Mechanistic Reasoning in Science and Technology Education?

The concept of mechanism goes back to the French philosopher Rene Descartes (1596–1650), who offered a machine-based vision of the universe by claiming that the universe works according to mechanical laws, and everything in the universe can be explained by reference to the arrangement and movements of its parts. Descartes's concept of a mechanical universe that came to be known as 'the mechanistic philosophy' implies that all happenings in the universe are governed by the laws of matter and motion alone, and their explanations can be reduced to their smallest constituent parts. The machine metaphor provides an insight into the behavior of many naturally occurring mechanisms, i.e., an account of how physical sciences and life sciences phenomena occur. This critical worldview contrasts with Aristotle's material substantial forms, in which he viewed knowledge as sense experience, i.e., all knowledge emanates from experience and evidence gathered by the five senses. Descartes was skeptical of Aristotle's empiricist view, as knowledge can also be derived through reasoning from innate ideas of the essences of things, independently of any sensory image.

Despite acceptance that mechanical explanations were the only intellectually satisfactory ones from early modern natural philosophers such as Boyle and Hooke, the concept was later criticized for being too broad and incomplete, but was never rejected altogether. For example, Newton's laws of motion have similitude to Descartes's laws of nature, but Newton's theory of universal gravitation is considered superior to Descartes's mechanical explanations of gravitation. Also, Descartes's mechanistic view of describing humans and animals as machines that function like complex automata was challenged mainly due to its difficulties in explaining the extreme complexities of living things in the mechanistic term. This criticism led to an active interest in the explanation of life phenomena among mechanistic philosophers. Subsequently, mechanical philosophy was as influential in the life sciences as it was in the physical sciences. Through time, the concept of mechanism has evolved in an attempt to respond to legitimate criticisms, and we suspect that it will continue to do so.

Contemporary mechanical philosophy that emerged around the turn of the twenty-first century has re-examined the traditional metaphysical categories of properties, relations and events for describing a phenomenon. The contemporary mechanical philosophy provides explanations of phenomena with reference to the elements of mechanisms and the linkages between them. Machamer et al. (2000) advocate that a satisfactory explanation in many fields of science requires describing a mechanism. They suggest, from a dualistic viewpoint, that *'the concept of mechanism is composed of entities and activities; activities are the producers of change and entities are things that are engaged in activities'* (Machamer et al., 2000, p.5). Entities are the material components of a mechanism and include the macroscopic (or concrete concepts) and microscopic (or abstract concepts) components. The entities are assumed to have distinctive implicit and explicit properties that influence the activities in which they are engaged. The implicit properties (e.g., elements, charges, bonding) are the recognizable features, whereas explicit properties (e.g., partial charges, potential energy, orbital size) are the non-recognizable features that characterize the entities. *'The entities and activities are organized in such a manner that they are responsible for the cause of the phenomenon'* (Machamer et al., 2000, p.5).

For describing a phenomenon as the behavior of the mechanism as a whole, Caspari and Graulich (Caspari & Graulich, 2019, p.110) suggest that the act of accounting for a phenomenon consists of: (1) identifying the entities of the mechanism and their properties; (2) identifying the activities in which the entities engage; (3) identifying the temporal and spatial organization of entities and activities; and (4) connecting those components to account for the dynamic sequence of mechanistic steps. This suggestion implies that the contemporary mechanical philosophy emphasizes that nature is hierarchically arranged and offers a structural and causal account of a phenomenon. Thus, the structural and causal account of a phenomenon provides an intelligible building block with increased complexity in reasoning.

Moreira et al. (2019) characterized four distinct levels of reasoning, namely: Level I (Descriptive), Level II (Relational), Level III (Simple Causal), and Level IV (Emerging Mechanistic), as shown in Fig. 17.1. The identification of major entities in a system and the recognition of their relevant properties (Level 1), to the construction of simple associations between entities and properties (Level 2), then to the identification of causal links mediated by interactions between entities and the activities to which they give place (Level 3) and finally to the identification of spatio-temporal organizations of entities that are involved in the activities (Level 4), depicts the increased level of complexity (sophistication) in reasoning (Moreira et al., 2019). Contemporary mechanical philosophy as a new framework for thinking about science and technology phenomena, therefore, provides a tool for students to think and reason scientifically and technologically.

As abstraction is ubiquitous in science and technology; providing an account of a phenomenon entails offering an explanation using scientific and technological concepts that exist on a spectrum ranging from simple to complex, and from concrete to abstract. We describe tangible concepts as 'concrete', while concepts that are inaccessible to the sensory and motor system are 'abstract'. Also, concepts have

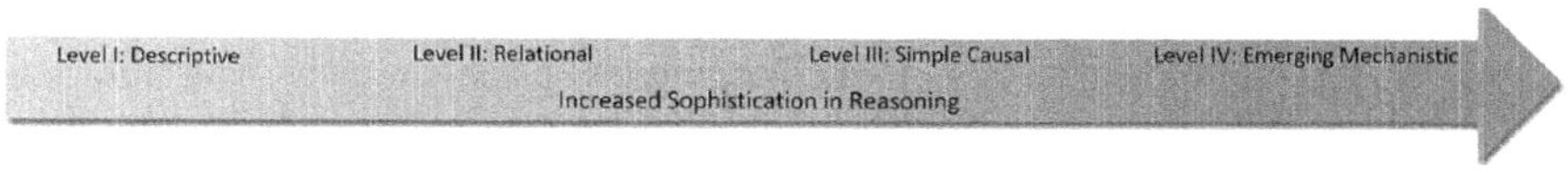

Fig. 17.1 Moreira et al.'s (2019) four distinct levels of reasoning

different levels of abstraction, as complex concepts (concepts with a greater degree of abstraction) are made up of intricate and interconnected ideas or concepts that require higher cognitive processes whereas simple concepts (concepts with a lower degree of abstraction) are the concepts that cannot be broken down any further into more basic or simpler concepts. Describing the mechanism for a phenomenon requires students to ask what constitutes the phenomenon under investigation, and how and why that particular phenomenon occurs. This implies that providing an account of a phenomenon requires students to examine concepts with a lower degree of abstraction (identifying the entities of the mechanism and their explicit properties) for building up an explanation that requires concepts with a greater degree of abstraction (connecting the different components to account for the dynamic sequence of mechanistic steps).

In view of the above, we assume that the varying degree of abstraction amongst concepts for describing a mechanism influences students' thinking and reasoning about the phenomenon. Reasoning about a mechanism is challenging for students because the depth and complexity of thinking and reasoning increase as the degree of abstraction increases i.e., the increasing level of complexity in reasoning correlates with the increasing degree of abstraction among the different concepts that are used for describing a specific mechanism. The distinct degree of abstraction, which includes concepts on a spectrum ranging from concrete to abstract and from simple to complex, is a major challenge for students to progress along the increasing level of complexity in reasoning when describing a mechanism that accounts for a phenomenon.

Science and technology educators can leverage emerging technologies as pedagogical resources to stimulate students' mechanistic reasoning for learning abstract and complex scientific and technological concepts. Studies have reported that emerging technologies can provide a virtual learning environment that extends learning beyond the physical space. Emerging technologies have the capacity to decrease the level of abstraction by transforming abstract and complex concepts into perceptible representations. In the following section, we will discuss how emerging technologies can stimulate students' mechanistic reasoning in science and technology education.

How Can Emerging Technologies Stimulate Students' Mechanistic Reasoning in Science and Technology Education?

Emerging technology refers to new technologies or technologies that are currently developing, or that are envisaged to be available in the near future as a result of successive innovations. Emerging educational technologies with varied characteristic features have a significant impact on the learning process by emphasizing how distinct information and knowledge can be presented and constructed. Each emerging educational technology has specific characteristic features that are more adaptable for a specific subject domain, as the nature of knowledge and subject specificity varies amongst subject domains. Characteristic features of emerging technologies, such as psychological immersion and interaction, have been well researched for providing a virtual learning environment in science and technology education. Yang and Baldwin (2020) investigate the technology-use strategies for supporting student learning in different integrated science, technology, engineering and mathematics (STEM) learning environments, and they state that expanding learning through immersive and interactive technology enhances learning. For further conceptual understanding of STEM education as a meta-discipline, please relate to Chap. 4.

Emerging technologies with immersive and interactive features enable students to interact with abstract concepts in a new and meaningful way through the process of reification, whereby abstract ideas or concepts are presented as inanimate objects. Emerging technologies with immersive and interactive features provide simulations and extend physical learning settings via computer technology (Yang & Baldwin, 2020). To describe a mechanism that necessitates an increasing level of complexity in reasoning due to the increasing level of abstraction in science and technology, emerging technologies with immersive and interactive features have an edge. This is because abstract and complex concepts can be made more concrete and experienceable when the emerging technologies have immersive and interactive features. Yet, to our knowledge, there is no discussion in the body of literature on how immersive and interactive features of technologies that provide a virtual learning environment have the potential to stimulate students' mechanistic reasoning in science and technology education.

In the following paragraphs, we attempt to examine how immersive and interactive features of emerging technologies have the potential to stimulate students' mechanistic reasoning by making the connection between the increasing level of complexity in reasoning due to the increasing level of abstraction, and the increasing level of psychological immersion and interaction when using emerging technologies.

Immersive Features of Emerging Technologies

Immersive features of emerging technologies allow students to visually comprehend abstract concepts in science and technology by animating what is invisible to the eye in a virtual environment. For example, one can view the explicit as well as the implicit properties of entities in a 3D environment and determine what entities are necessary for an activity to occur, or how the properties of the entities influence its activity. Immersive features of technology provide students with a physical presence when dealing with abstract ideas and concepts (or virtual objects and processes), which can be experienced as actual physical objects artificially created or simulated. Physical presence refers to the experience or feeling of being present in a virtual environment, rather than the immediate physical environment. Subsequently, one has sensory experiences of the connection between the different elements of a mechanism with reference to its identity, activity and organization when describing a mechanism for explaining how and why a natural and technological phenomenon occurs. When dealing with abstract concepts, studies have suggested that senses stimulated in a virtual classroom have the potential to significantly enhance a user's sense of presence or psychological immersion.

The art of explanation in science and technology education lies in providing enough concrete details to help students understand a phenomenon. When abstract objects and processes are presented concretely in a virtual environment, students are more engaged as the physical presence increases students' intrinsic motivation. Stimulating students' intrinsic motivation to learn is a more effective strategy to get and keep students interested and engaged. For further details on science and technology teaching strategies, please refer to Chap. 15. As physical presence or psychological immersion is strongly required to visualize many kinds of information in a wide variety of graphical and animated forms, students will be in a better situation to overcome the challenges of engaging with abstract and complex concepts. Thus, decreasing the level of abstraction in science and technology by providing students with the physical presence in a virtual environment enables students to progress easier along the increasing level of complexity in reasoning when describing the mechanism of a phenomenon. Therefore, emerging technologies with an elevated level of psychological immersion have the potential to stimulate students' mechanistic reasoning in science and technology education, by providing them with a physical presence in a virtual environment that enhances students' engagement with abstract and complex concepts in the learning process.

Interactive Features of Emerging Technologies

Another characteristic feature of emerging technologies is their ability to enable students to operate or intervene in a virtual environment by interacting with objects without having them in their hands. The maneuverability of objects in the virtual environment requires human-computer interaction with scenes and objects that appear to be real, thereby making the content interactive. The implicit properties of entities, their activities and their organisations are abstract and complex concepts that can appear as scenes and objects in a virtual environment. When students manipulate the entities of a mechanism as virtual objects, they can immediately describe the cause-and-effect relationship between the entities. Digital 3D content can be designed to enable students to engage with entities of a mechanism and explore the different implicit properties of the entities as well as their activities, by interacting with the objects that appear to be real. Subsequently, with interactive features of emerging technologies, students can have hands-on experiences that allow them to learn by interacting (doing). For a conceptual understanding of hands-on experience, please refer to Chap. 9 on real and virtual laboratories.

Learning by interacting in a virtual environment allows students to take control of their experiences and actions during the construction of the explanation of a phenomenon. Emerging technologies with interactive features require more sophisticated virtual environments that give students the options for agency within the environment. The agency refers to the experiences of autonomy and control over one's actions during the learning process. For example, Wang et al. (2022) designed a game, *Cellverse,* in a virtual environment for students to have the agency of exploring the environment, selecting organelles, learning more about them by opening up the clipboard, and collecting samples of possible evidence for the type of cystic fibrosis in the game. Emerging technologies with a higher level of interactivity provide students with a higher level of autonomy and control, or a higher level of agency, in their learning. Thus, students will be in a better position to identify the temporal and spatial organization of entities and activities, and to connect those components to account for the dynamic sequence of mechanistic steps. Therefore, emerging technologies with interactive features can stimulate students' mechanistic reasoning in science and technology education by providing them with an agency in a virtual environment that enhances students' exploration and explanation.

Emerging Technologies to Stimulate Students' Mechanistic Reasoning in Science and Technology Education

Extended reality (XR) is the umbrella term for describing all kinds of altered realities generated by computer technology and wearables that merge the physical and virtual worlds or create an entirely virtual experience. XR technologies can extend the reality that people experience by blending the real and virtual worlds and extend across the virtual reality (VR), augmented reality (AR) and mixed reality (MR) (Morimoto et al., 2022), as well as all future technologies that refer to all real-and-virtual combined environments and human-machine interactions. All three 'realities' share common overlapping features and requirements, but have marked differences.

To better comprehend the marked differences, we briefly outline in the following paragraphs the three realities (VR, AR and MR) in terms of Milgram et al.'s (1995) reality-virtuality continuum, which covers the range of possibilities between the entirely physical world and the fully digital world. The marked differences will then be aligned with the levels of immersive and interactive features of emerging technologies for stimulating students' mechanistic reasoning. We have previously assumed that immersive and interactive features of technology enable teachers to overcome the challenges of increasing levels of complexity in students' reasoning, which arise due to the increasing degree of abstraction amongst different concepts for constructing an explanation about a scientific and technological phenomenon. Therefore, this enables us to understand which emerging technologies can be used to stimulate students' mechanistic reasoning in science and technology education.

Milgram et al. (1995) considered the real environment and the virtual environment as a continuum, whereby the real environment and the virtual environment are taken as two ends of the continuum respectively, as shown in Fig. 17.2. The reality-virtuality continuum is a continuous scale, ranging between the completely real and the completely virtual. The reality-virtuality continuum then encompasses all variations and compositions of real and virtual objects, ideas and concepts. The left end (or the starting point) of the continuum shows a real environment that inwardly extends to form AR, while the right end (or the endpoint) of the virtual environment is extended to form augmented virtuality (Milgram et al., 1995). MR is located between the real environment and the virtual environment.

With VR technologies, the students are part of a fully digital, simulated environment that they may not otherwise experience. VR may draw from actual reality, but it is its own version, without any grounding in the world around us, as it represents the end point of the reality-virtuality continuum. Special VR headsets are used to fully immerse students in virtual reality, thereby providing students with a greater sense of presence. The VR hand-held controllers provide better interaction and manipulation with the digital content depending on their degrees of freedom, thus providing students with a greater sense of agency.

Meanwhile, AR is an interactive experience of a real-world environment whereby virtual information and objects are overlaid in the real world, thus superimposing added information on an otherwise unaltered reality. It can be used to display definitions, models, or project movement with the end goal of adding visual elements to students' perception of reality, as it represents the starting point of the reality-virtuality continuum. Unlike VR headsets, AR glasses and headsets do not immerse users in a fully virtual environment, but just add digital objects to the real world. Also, AR provides superficial interaction between the digital elements and the physical world elements.

In between the false reality of VR and the absolute reality of AR is MR, as the latter lies in the middle of the reality-virtuality continuum. MR has recently garnered attention, mitigating the limitations of VR's exclusion of the real-world environment and AR's inability to interact with 3D data packets (Morimoto et al., 2022). MR provides a mixed-reality environment that allows digital and real-world objects to exist together and interact in real-time. With MR, students can see virtual objects just like with AR, but these objects can also interact with the real world. In a sense, MR is a more immersive and interactive type of AR.

The immersive and interactive features of VR, AR and MR vary along with the reality-virtuality continuum. To situate the technologies according to the degree of immersion, Tremosa (2022) adapts the simplified representation of Milgram et al.'s (1995) reality-virtuality continuum by incorporating XR technologies into the continuum as well as their degree of immersion. The representation shows that AR has

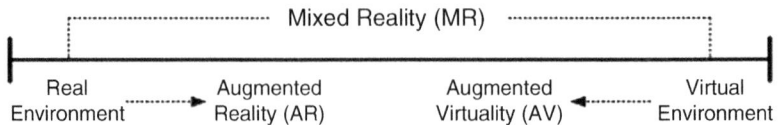

Fig. 17.2 Milgram et al.'s (1995) representation of the reality-virtuality continuum

a lower level of immersion compared to VR. We extend Tremosa's (2022) representation of current XR technologies by including the element of interaction. Similar to immersion, the level of interaction is low for AR and high for VR, as shown in Fig. 17.3.

VR is used for educational experiences that let the students escape their surroundings and immerse themselves in an entirely different setting. This entirely different setting allows learners to interact with abstract objects, concepts and processes. For example, Mintz et al. (2001) presented a new VR environment that employs a dynamic 3D model of the solar system based on powerful scientific visualization techniques. They reported that the 3D model allows students to enter a virtual model of the physical world, journey through it, zoom in or out as they wish, and change their viewpoints and perspectives, as the virtual world continues to behave and operate in its usual manner. This implies that students' level of psychological immersion and interaction is maximum at the endpoint of the virtuality continuum, as the immersive and interactive features of VR, AR and MR vary along with the reality-virtuality continuum. VR technologies, having in-depth immersive and interactive features, provide students with an engaging and explorational learning experience, which, in science and technology education, is crucial for an in-depth understanding of complex and abstract concepts.

With VR technologies, students are immersed in and interact with complex and abstract concepts that increase their engagement in the learning process and allow them to explore and construct an explanation of a scientific and technological phenomenon. Petersen et al. (2022) examine the process of learning with VR technology and suggest that immersive and interactive features of VR technology facilitate learning via presence and agency respectively. Presence and agency, when reasoning about a mechanism, enable students to progress up to Moreira et al.'s (2019)

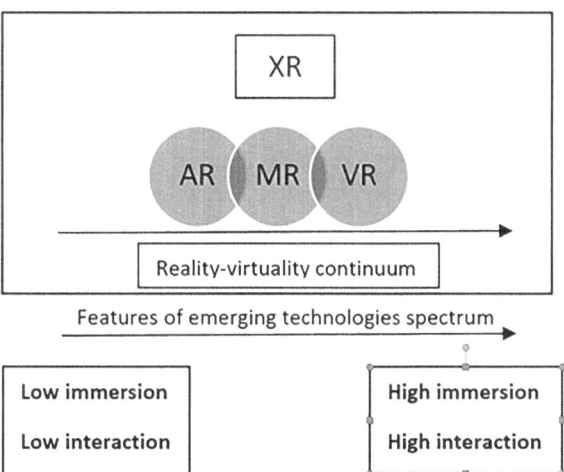

Fig. 17.3 Extended representation of the reality-virtuality continuum (adapted from Tremosa, 2022)

Level 4 of the increased complexity in reasoning (see Fig. 17.1), as discussed in the previous section. Creating immersive and interactive VR learning and teaching experiences in science and technology, which can be complementary to more traditional learning and teaching lectures, practicals and field methods, has the full potential to stimulate students' mechanistic reasoning.

Conclusion

Science and technology education must be improved to better equip today's students with the competencies for the job market of tomorrow, due to major shifts in employment practices. Developing students' mechanistic reasoning, which demands the provision of a structural and causal account of a mechanism, is critical to disciplined inquiry in the science and technology domain, but is quite challenging. It is challenging because students have to progress through an increasing level of complexity in reasoning when describing a natural and technological phenomenon (Moreira et al., 2019). As abstraction is ubiquitous in science and technology, describing a natural and technological phenomenon requires students to use and understand scientific and technological concepts that exist on a spectrum ranging from simple to complex, and from concrete to abstract. The increasing level of complexity in reasoning for describing a phenomenon is due to the increasing level of abstraction in scientific and technological concepts. Educators should leverage emerging technologies to overcome the challenge of developing students' mechanistic reasoning.

Technologies provide a reality-virtuality continuum that connects completely real environments to completely virtual ones (Milgram et al., 1995). Emerging technologies on the virtual extremum translate abstract objects and relations, typically represented in textual forms, into animated representations in a virtual environment. Yang and Baldwin (2020) suggest that immersive and interactive features of emerging technologies enhance STEM learning, and Petersen et al. (2022) claim that immersive and interactive features of technology facilitate learning via presence and agency. Tremosa (2022) presents the current XR technologies according to the spectrum of immersion, which we, therefore, extend by including the spectrum of interaction as both spectra are congruent with each other due to their linear relationship. An increasing degree of immersive and interactive features of emerging technologies enhances learning.

In this chapter, we use existing findings in the research literature to provide a discourse about how emerging technologies have the potential to stimulate students' mechanistic reasoning in science and technology education. We adopt Moreira et al.'s (2019) mechanistic framework and adapt Tremosa's (2022) representation of the current XR technologies to make the connection between the increasing level of complexity in reasoning due to the increasing level of abstraction, and the increasing level of psychological immersion and interaction when using emerging

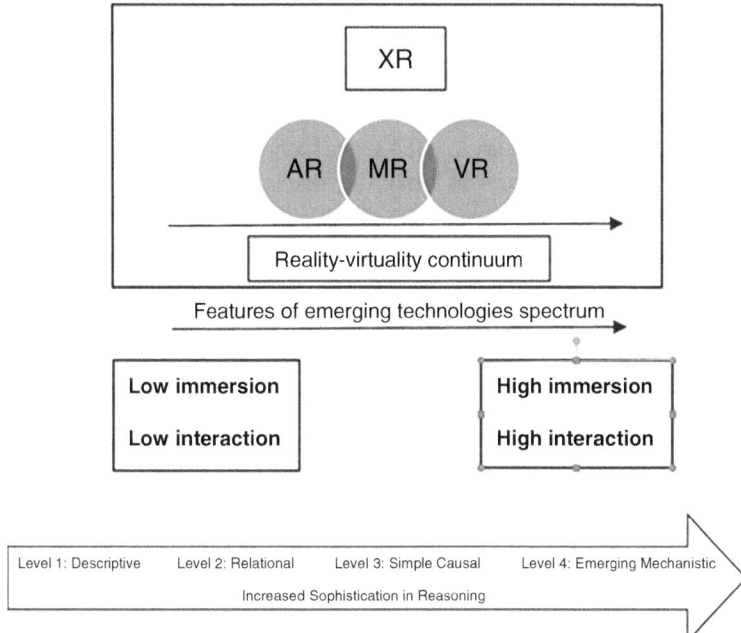

Fig. 17.4 Representation of the reality-virtuality continuum for stimulating students' mechanistic reasoning

technologies. Therefore, we represent the discourse in the form of a conceptual structure (Fig. 17.4) that will enable readers to clarify their comprehension of this chapter.

Emergent educational technologies such as virtual reality (VR) and augmented reality (AR) have the potential to revolutionize education by supporting improvement through innovative practices. With state-of-the-art technology, new immersive and interactive learning environments are explored to provide more effective and efficient ways of enhancing students' understanding of natural and technological phenomena when compared to traditional teaching. At present, however, only a very few virtual reality software packages have been developed.

Recommendations

The above discussions have highlighted the need to develop learners' mechanistic reasoning to enhance science and technology education, prepare them to operate successfully as twenty-first century citizens, and perform effectively as workers in new jobs that will be created in the future with the advent of industry 4.0. It has also been argued that emerging technologies can support the development of mechanistic reasoning in science and technology education. Thus, the following recommendations are proposed for the development of mechanistic reasoning in science and technology education through the use of emerging technologies.

- Educators and educational researchers must engage in empirical research on the use of emerging technologies in science and technology education, more specifically on how these can best be used to promote the development of learners' mechanistic reasoning. This would provide evidence-based findings on the effectiveness of using emerging technologies for promoting mechanistic reasoning in science and technology.
- Teacher education and continuous professional development (CPD) programs need to incorporate emerging technologies and mechanistic reasoning as core components in their science and technology education curricula for both in-service and pre-service educators. These programs must also endow the trainee educators with the competencies to use emerging technologies for developing mechanistic reasoning amongst learners.
- Policymakers and other education-related stakeholders must ensure that necessary logistics, resources and capacity-building are provided to facilitate the use of emerging technologies by educators in science and technology education.
- Science and technology educators must work synergistically with software developers to ensure that digital educational resources based on emerging technologies (such as VR, AR and MR) being developed are in line with the relevant curricula and are developmentally appropriate for the targeted learners.
- Emerging technologies should be used through inquiry and the conceptual change approach to promote the development of critical thinking, problem-solving, and creative thinking skills, which are essential for promoting mechanistic reasoning among learners. Furthermore, the conceptual change approach would help to address learners' misconceptions that could hinder effective learning.
- Educational programs need to ensure that learners develop the necessary competencies for using emerging technologies to support and enhance their learning.

Summary

In this chapter, the importance of mechanistic reasoning in science and technology education has been highlighted, especially in understanding and explaining the cause-effect relationship of a phenomenon. The importance of emerging technologies in stimulating students' mechanistic reasoning in science and technology has been elaborated. This is attributed to the fact that the immersive and interactive features of the emerging technologies can help in making abstract and complex science and technology concepts more concrete and accessible to learners. The chapter has also elaborated on how learning in a virtual environment (provided by emerging technologies) can enhance students' engagement with abstract and complex concepts helping them to construct explanation of different phenomena, thereby promoting the development of mechanistic reasoning.

Further Reading

Mechanistic science https://www.ncbi.nlm.nih.gov/pmc/articles/PMC2738038/
Six theses on mechanisms and mechanistic science https://link.springer.com/article/10.1007/s10838-021-09587-x

References

Bolger, M., Kobiela, M., Weinberg, P. & Lehrer, R. (2009, June). Analysis of children's mechanistic reasoning about linkages and levers in the context of engineering design. In *2009 annual conference & exposition* (pp. 14–214).

Caspari, I., & Graulich, N. (2019). Scaffolding the structure of organic chemistry students' multivariate comparative mechanistic reasoning. *International Journal of Physics & Chemistry Education, 11*(2), 31–43.

De Andrade, V., Shwartz, Y., Freire, S., & Baptista, M. (2022). Students' mechanistic reasoning in practice: Enabling functions of drawing, gestures and talk. *Science Education, 106*(1), 199–225.

Kurkul, K. E., Castine, E., Leech, K., & Corriveau, K. H. (2021). How does a switch work? The relation between adult mechanistic language and children's learning. *Journal of Applied Developmental Psychology, 72*, 101221.

Machamer, P., Darden, L., & Craver, C. F. (2000). Thinking about mechanisms. *Philosophy of Science, 67*(1), 1–25.

Moreira, P., Marzabal, A., & Talanquer, V. (2019). Using a mechanistic framework to characterise chemistry students' reasoning in written explanations. *Chemistry Education Research and Practice, 20*(1), 120–131.

Morimoto, T., Kobayashi, T., Hirata, H., Otani, K., Sugimoto, M., Tsukamoto, M., & Mawatari, M. (2022). XR (extended reality: Virtual reality, augmented reality, mixed reality) technology in spine medicine: *Status quo* and *quo vadis. Journal of Clinical Medicine, 11*(2), 470.

Milgram, P., Takemura, H., Utsumi, A. & Kishino, F. (1995, December). Augmented reality: A class of displays on the reality-virtuality continuum. *Telemanipulator and Telepresence Technologies, 2351*, 282–292.

Mintz, R., Litvak, S., & Yair, Y. (2001). 3D-virtual reality in science education: an implication for astronomy teaching. *Journal of Computers in Mathematics and Science Teaching, 20*(3), 293–305.

Petersen, G. B., Petkakis, G., & Makransky, G. (2022). A study of how immersion and interactivity drive VR learning. *Computers & Education, 179*, 104429.

Tremosa, L. (2022, January 26). *Beyond AR vs. VR: What is the difference between AR vs. MR vs. VR vs. XR? Interaction design foundation (IxDF).* Retrieved from: https://www.interaction-design.org/literature/article/beyond-ar-vs-vr-what-is-the-difference-between-ar-vs-mr-vs-vr-vs-xr Accessed 09 July 22.

Wang, A., Thompson, M., Roy, D., Pan, K., Perry, J., Tan, P., & Klopfer, E. (2022). Iterative user and expert feedback in the design of an educational virtual reality biology game. *Interactive Learning Environments, 30*(4), 677–694.

Yang, D., & Baldwin, S. J. (2020). Using technology to support student learning in an integrated STEM learning environment. *International Journal of Technology in Education and Science, 4*(1), 1–11.

Vickren Narrainsawmy joined the Mauritius Institute of Education in June 2018. He joined the Institution after five years of experience as a secondary school educator and holds a Master's degree in Chemistry (specialization in Organic Chemistry) and in Education. He has about 4 years of experience in teacher education, curriculum development and educational research in science education and chemistry education. His research interests include STEM education, inquiry learning, remote teaching and mechanistic reasoning.

Dr Fawzia Narod is Associate Professor in the Department of Science Education at the Mauritius Institute of Education. She has many years' experience in teacher education and development of curriculum and curriculum materials at the pre-primary, primary and secondary levels, with particular focus on science education and chemistry education. In addition to teaching and coordination of teacher education courses and programmes, she is actively engaged in educational research, and the supervision of undergraduate and postgraduate research dissertations. She has been the Lead Investigator, and an active member, of several research projects. Dr Narod also supervises MA and PhD research dissertations for the University of Brighton (UK) and University of KwaZulu Natal (South Africa). Her research interests include STEM education, science education, chemistry education, use of ICT as a pedagogical tool, mechanistic reasoning, teacher education and higher education, remote teaching and online teaching and learning. She has authored several articles in peer-reviewed journals, peer-reviewed conference proceedings and book chapters on science education and chemistry education.

Chapter 18
Problem-Solving in Science and Technology Education

Bulent Çavaş, Pınar Çavaş, and Yasemin Özdem Yılmaz

Abstract This chapter focuses on problem-solving, which involves describing a problem, figuring out its root cause, locating, ranking and choosing potential solutions, as well as putting those solutions into action in science and technology education. This chapter covers (1) what problem-solving means for science and technology education; (2) what the problem-solving processes are and how these processes can be used step-by-step for effective problem-solving and (3) the use of problem-solving in citizen science projects supported by the European Union. The chapter also includes discussion of and recommendations for future scientific research in the field of science and technology education.

Keywords Problem-solving · Processes · Citizen science · Science and technology education

Introduction

In the changing and developing world, with what kind of knowledge, skills and competencies the new generation learners will be equipped is one of the most crucial questions that science educators seek to answer. What skills and information do our children need to withstand the rapid changes that appear to be occurring in every

B. Çavaş (✉)
Faculty of Education, Dokuz Eylül University, Buca, Izmir, Türkiye
e-mail: bulent.cavas@deu.edu.tr

P. Çavaş
Faculty of Education, Ege University, Bornova, Izmir, Türkiye
e-mail: pinar.cavas@ege.edu.tr

Y. Ö. Yılmaz
Muğla Sıtkı Koçman University, Faculty of Education, Muğla, Türkiye
e-mail: yaseminozdem@mu.edu.tr

© The Author(s), under exclusive license to Springer Nature Switzerland AG 2023
B. Akpan et al. (eds.), *Contemporary Issues in Science and Technology Education*, Contemporary Trends and Issues in Science Education 56, https://doi.org/10.1007/978-3-031-24259-5_18

aspect of life? Our children's knowledge and abilities will be out of date when they need to use them in their personal lives and the workplace, if we prepare them for current opportunities. The students should acquire some critical skills that will enable them to survive and adapt to a new world. These skills are mostly called twenty-first century skills, and problem-solving is one of the more critical skills (for other skills, please see Sect. III of this book). Binkley et al. (2012) put twenty-first century skills into four main categories: ways of thinking, ways of working, tools for working, and skills for living in the world. Problem-solving is related to thinking skills and classified into ways of thinking skills.

Problem-solving is seen as an essential process in facilitating students' learning and the acquisition of many metacognitive skills in science and technology education (Garrett, 1986). Problem-solving skills are a process, including usage of previously gained knowledge, skills and understanding, to satisfy the demands of an unknown situation. The procedure begins with the initial confrontation and ends once a response has been acquired and taken into account in light of the initial circumstances. The student must synthesize what he or she has learned, and apply it to the new and different situation.

Science education provides training for both scientists and technologists towards the advancement of the nation's technical and economic capabilities. For that reason, it is very important to provide a quality science and technology education from an early age. Problem-solving skills are needed for the teaching and learning of science in order to help students develop their ability to solve problems in science and technology. It was claimed that solving problems constituted a unique kind of meaningful learning (Ausubel, 1968). This may be connected to the idea that problem-solving is *'the mental process employed in arriving at a "best" solution to an unknown subject to a series of unknowns'* (Woods, 1987). On the other hand, according to some literature, although students have the subject knowledge, they still lack the ability to use it to solve issues in science and technology (Chi, Feltovich & Glaser, 1981; Hobden, 1998; Osborne & Dillon, 2008). According to Taconis (1995), this is likely to be a result of students not being taught how to use their own problem-solving skills.

The following sections of this chapter contain information on how to improve students' problem-solving skills in science and technology learning and teaching environments.

The Problem-Solving Process

Many researchers proposed phases or steps to simplify the problem-solving process. However, according to Anderson (1967), problem-solving is an intuitive process, which is later checked analytically. He cites Bruner (1962, cited in Anderson, 1967) in saying that rather than using set formulas or patterns, intuitive problem-solving appears to be based on an implicit awareness of the entire issue, and also makes use of leaps, skips and shortcuts. Moreover, while the problem-solving process is creative, the steps limit the variability of methods and overemphasize, following a pattern (Anderson, 1967). Brownell (1942, cited in Anderson, 1967) criticizes the steps for (1) not being an evidence-based demonstration of good thinking; and (2) ignoring other elements that could affect problem-solving. In the nature of science, there may be as many scientific methods as there are problems and scientists to solve them (Hurd, 1960, cited in Anderson, 1967; McComas, 1998).

Agreeing with the criticisms, therefore, the steps given below do not represent a fixed procedure to follow. Rather, they are the processes frequently encountered in problem-solving. These frequently-used processes are briefly described in the problem-solving framework established by the PISA consortium as: '*Problem-solving begins with recognising that a problem situation exists and establishing an understanding of the nature of the situation. It requires the solver to identify the specific problem(s) to be solved and to plan and carry out a solution, along with monitoring and evaluating progress throughout the activity*' (OECD, 2013, p.123).

Identifying the Problem

Lawson (2003) recognizes the problem as a situation where the methods or procedures leading to answers or solutions to reach the desired goal are not readily available. Not all problems are clearly structured, include all necessary information, or have a single solution. As a matter of fact, most problems in science and technology have an ill-defined structure (Greenwald, 2000). Ill-defined problems are less clearly structured, may lack a single, accepted solution, and may not contain all pertinent information for reaching a solution (Chin & Chia, 2006; Lawson, 2003; Simon, 1973). Gallagher *et al.* (1995) also argue that ill-structured problems are interdisciplinary, which enables problem-solvers to see numerous, diverse examples of how various disciplines approach a particular issue and collaborate when addressing problems (Chin & Chia, 2006).

In schools, problems are frequently well-structured, well-defined, self-contained, and have just one correct response (Gallagher, Stepien & Rosenthal, 1992). In reality, however, many problems of science and technology are ill-structured. Jonassen (1997) indicates that ill-structured problem-solving is more like a design process, rather than a methodical search for solutions to problems. He claims that identifying if the problem actually exists is the first stage in the problem-solving process, because the ill-structured problem may not appear directly, or may be hidden. Moreover, there may be several representations or understandings of the problem, which is another characteristic making it ill-structured (Jonassen, 1997). Therefore, identifying a suitable problem to work on from the competing possibilities is important in ill-structured problem-solving.

Contextualizing a problem is a way to help to identify it (Mahanal, 2022). The role of context in problem-solving is also largely supported by research in cognitive science (Glaser, 1992). For example, Murphy and McCormick (1997) argue that '*To ensure a task, that is set as a problem, is personally meaningful, students must be involved in the context of the problem, that is the embedding features of the problem; for example, in technology the designing and making of an aid for a disabled child in a special school, and in science, exploring the effects of different wheel treads on distance travelled by a vehicle in the context of safety and travel'* (p.463).

Contextualizing is especially useful in technology-related problems; however, science does not always deal with contextualized problems. Murphy and McCormick (1997) indicate that describing the role of context in science may be more difficult, because adding context to an issue makes it more technological, which, according to studies, makes it more challenging for students to understand the science behind the problem. In these situations, they suggest that the problem be directly and authentically tied to the context, yet the problem itself must be a problem that science can solve.

Analyzing the Problem

A problem analysis is the breakdown of the problem in order for the problem-solver to better comprehend it and suggest workable options for solving it. In other words, the goal of problem analysis is to gain a better understanding of the problem being solved before generating a solution.

According to Osborn (1953), analysis is essential, especially when defining the goals and narrowing the scope of our aims. For example, separating the portions of a problem that require ideas from those that require judgement is one way to use analysis. By doing this, he claims, we might avoid the confusion that can occasionally prevent creative thinking. Moreover, Osborn (1953) indicates that analysis by itself can reveal hints that quicken our capacity for association and so fuel our imagination.

The two fundamental tasks of analysis are learning more about the problem circumstance and clarifying whatever you already know about it (Koberg & Bagnall, 1981). The IDEAL problem-solving approach developed by Bransford and Stein (1984) describes analysis as a very careful and systematic approach. During analysis, problem-solvers usually divide the defined problem into smaller, easier-to-solve problems. Bransford and Stein (1984) claim that to deconstruct complex issues into component elements in order to succeed is a natural human reaction to problems. When the problem is divided into its component elements, it becomes relatively straightforward.

Describing the Problem

Describing the problem is another important step towards solving a problem (von Hippel & von Krogh, 2016). Describing problems helps students, since this enables them to track down problems in a systematic manner. The amount, quality, originality and type of solutions offered are all substantially influenced by how clearly specified the problem is (Mahanal *et al.,* 2022). When the problem is well-described, Presseisen (1985) argues, there is a greater likelihood that the students will identify common pattern systems and be inspired to solve problems effectively.

Developing Alternate Solutions

The problem-solver is required to consider alternative viewpoints when confronting an ill-structured problem and develop justifications for the suggested solution. In science, alternative hypotheses are constructed for this purpose. According to Osborn (1953), our progress toward the resolution of an ill-structured problem is likely to depend on the degree to which we accumulate hypotheses once we have stated our goals and gathered sufficient data. The advantage of generating hypotheses is that the more concepts we formulate as potential solutions, the more probable it is that we will discover the concept or concepts that will address the problem. Osborn (1953) claims that our tentative ideas and hypotheses can turn out to be the very solutions that we need. Moreover, our thoughts are more likely to spark additional ideas, which could then lead to the discovery of solutions. Additionally, it is common in science for each new theory to suggest a direction for additional research, which could ultimately lead to a solution.

Similarly, examining potential solutions to a problem is the third step in the IDEAL problem-solving process (Bransford & Stein, 1984). This frequently entails re-evaluating your objectives as well as taking into account potential solutions or tactics that could be used to accomplish those objectives. Mahanal *et al.* (2022) suggest strategies such as brainstorming, surveys and discussion groups to generate alternative solutions.

Implementing the Solution

The option that was selected as being the one most capable of achieving an anticipated solution is finally put to the test during implementation. Problem-solving creativity is necessary for implementing a solution. The methods chosen, the data gathered and the tools used for data gathering are all parts of thinking during the implementation of a solution. As Ioannidou and Erduran (2021) put forward, scientific evidence is produced by scientists using a wide range of techniques and resources. The type of the problem that scientists consider and the instruments and procedures that are available over a certain period are frequently factors in the method that is chosen.

According to Nezu (2004), the goal of solution implementation and verification is to put the solution plan into action, to monitor and assess its performance, and to troubleshoot if the result is unsatisfactory. To put it another way, if the solution doesn't work, the problem-solver needs to go through the different issue-solving activities again to work out where additional efforts should be focused in order to address the problem effectively.

Collecting and Analyzing the Data

Data are measurements and observations made in the course of science that, after being examined and interpreted, can be used to support a claim. All scientific research relies on data, which are gathered in some way by a scientist. Students can collect data through their laboratory work in real or in virtual settings as described by Cassam Atchia and Rumjaun (Chap. 9 in this book) or by observing nature. However, scientific knowledge is considerably more than just a simple compiling of data points, and gathering data is simply one step in scientific research. All scientists must decide which data are most pertinent to their research and what to do with them, including how to process and analyze a collection of measurements to create a meaningful dataset, and how to interpret the information in light of their prior knowledge (Egger & Carpi, 2008). Data can be converted into evidence-supporting scientific concepts, arguments and hypotheses through careful and methodical gathering, analysis and interpretation (Sampson, Enderle & Grooms, 2013).

The process of understanding the significance of the data gathered, arranged, and shown in the form of a table or graph is known as data analysis. Searching for patterns – similarities, differences, trends and other relationships – and considering what these patterns might indicate are both steps in the process. Scientists, at last, conclude by summarizing their results and connecting them to their original hypothesis.

Given the similarities between solving a problem and conducting an investigation in science, Pérez and Torregrosa (1983) assert that the analysis of data and interpretation of outcomes are noteworthy in the problem-solving process. They argue that results of the analysis allow us to evaluate the validity of the hypotheses put forward, as well as the degree to which the qualitative assessment of the situation made at the beginning was accurate, and the suitability of the adopted tactics.

Sharing the Results

The Search, Solve, Create and Share (SSCS) model created by Pizzini (1989) explains that the Share phase's main goal is to get students talking about how they solved problems or answered questions. The Share phase does not necessarily take place at the end of the problem-solving process, but it often centers on the finished product. According to Pizzini (1989), the share phase involves more than just talking to students and other people. In this phase, students express their thoughts through dialogue and engagement, absorb feedback, consider and assess solutions and responses, and come up with possible new research questions. In other words, this phase requires collaboration as much as communication as described by Odell, Dyer and Klett (Chap. 20 in this book). When an accepted solution produces a new problem, or when flawed logic or mistakes in the problem-solving strategy are found through external examination of the shared product, new potential Search questions are generated. This, therefore, enables the problem-solver to pinpoint problem-solving techniques that require improvement and to come up with fresh Search questions.

Problem-Solving Examples in European Union Projects

The European Union has developed policies to support the participation of more citizens in scientific research processes. As a result of these policies, project calls were published to ensure the active participation of citizens in scientific research processes. Citizens' participation in these projects has been developed under various roles, such as observer, data provider and data analyst. Through the participation of citizens in scientific research processes, it is expected both that scientific literacy levels will increase across Europe, and quality scientific research results will be obtained. All the projects examined under this process are within the scope of the European Union's most prestigious research program, the framework program entitled *Horizon 2020,* and also the Erasmus Plus program. When we look at the content of these projects, we see that problem-solving skills are at the forefront. In this context, the projects examined below include the problem-solving skills of citizens in particular.

Achieving a New European Energy Awareness (AURORA)

The AURORA Project is one of the projects targeting climate change and encouraging less energy use. In this project, a mobile application has been developed, with the participation of approximately 7000 citizens from Denmark, England, Portugal, Slovenia and Spain, where citizens can decide on their energy use. In this project, a joint initiative is planned to change the energy use behavior of citizens, to pay for less energy and to emit less carbon, through using their problem-solving skills. In addition, citizens are encouraged to invest more in renewable energy sources.

The project's web address is https://www.aurora-h2020.eu/, where detailed information about AURORA can be found.

Creating School Seismology Labs for the Development of Students'
Competences (SEISMO-Lab)

The SEISMO-Lab project is an important project that focuses on earthquake issues and is supported by the Erasmus Plus program and co-ordinated by the National Observatory of Athens in Greece. This is a project designed to enable students to access more detailed information about earthquakes, with the help of seismometers in their schools, by using their problem-solving skills. Students are asked to find the locations of earthquakes using real earthquake data. The students would work like seismologists and find the epicenter of the earthquake by using earthquake data, develop an early warning system, and create the sound of the earthquake by converting the earthquake data into sound waves. The project also benefits from the important material generated by a previous Erasmus plus project called SNAC. The SNAC book entitled *Recommendations for Future Use,* developed within the framework of the project, describes in detail how students can 'play' with earthquake data by using their problem-solving skills (Milopoulos & Cerri, 2020).

The project's web address is http://snac.gein.noa.gr/, where more detailed information can be found.

CitizensHack2022

The goal of citizen engagement in research and innovation (R&I) is to enable individuals to participate in R&I activities, for instance through jointly generating innovative solutions to local problems. The necessity for co-operation, information-sharing and quick application of R&I findings to provide solutions that matter to citizens has never been more important than it is today. A culture of openness, inclusion and trust is fostered by new forms of collaboration between scientists, entrepreneurs and community members, improving the value of science to society and shaping policy decisions. CitizensHack2022 tries a novel citizen involvement strategy for converting R&I findings into societal benefit. It allows citizens to participate as active community members and co-create answers to the problems that they confront (through new business models, social innovations, prototypes, tests, proven concepts, demos, etc.). The project advances public understanding of science and technology, as discussed by Yingprayoon (Chap. 13 in this book). This is accomplished by collaborating throughout the hack with researchers and creators who use scientific data and research to inform their work.

The project's web address is https://ultrahack.org/citizenshack-2022, where more detailed information can be found.

Science That Makes Me Move (SMOVE)

In a school-based environment, researchers and students (classes 8 (age 13) and above) collaborate on a cross-sectional study assessing physical activity and sedentary behavior in students in Berlin and Brandenburg over the course of 1 week, using the activPALTM accelerometer and identifying factors (behavioral, socio-economic, environmental and others) that are associated with physical activity and sedentary behavior. The students create a questionnaire to evaluate factors that they consider of potential influence on their physical activity and sedentary behavior in collaboration with scientists from the Max-Delbrück-Center for Molecular Medicine in the Helmholtz Association (MDC), in order to identify these factors. Additionally, established student, parent and teacher surveys are employed. MDC scientists visit the participating schools up to three times during the field phase. Written acknowledgment of participation is given and the students are informed about the value to health of leading an active lifestyle during the scientists' initial visit to the classrooms. The students work in groups to construct a questionnaire to gauge what they believe to be potential influencing elements of physical activity and sedentary behavior. During the second visit by the MDC scientists, the students fill out questionnaires (newly-developed class-specific, plus established, questionnaires) and wear the activPALTM accelerometer, a small device that objectively records the students' movements from acceleration, over a 1-week period. Data are collected, anonymized, and prepared for in-class analysis and interactive interpretation during the third MDC visit.

The project's web address is https://www.mdc-berlin.de/content/smove-science-makes-me-move, where more detailed information can be found.

IANUS Peacelab

A facility for citizen science peace research is called IANUS Peacelab. Models, prototypes and apps are created and evaluated in the cognitive space. Citizens collaborate in a problem-oriented way, without separating social science, natural science, humanities, or engineering techniques according to their respective discipline boundaries, making it a laboratory. Make Peace, Not War – peace must also be constructed and put together. This goes beyond the issue of arms control. Vaccines, luxury items, food, raw minerals and information and communication technology all have the potential to either cause war or promote peaceful co-existence.

More information can be found at ishttps://eu-citizen.science/project/238.

Discussion and Recommendations

We thought that it would be useful to consider the following points related to problem-solving in future studies in science and technology education:

- In order to use problem-solving in teaching and learning environments, teachers' knowledge and skills should be checked. It is unlikely that teachers who do not have sufficient knowledge and skills about problem-solving processes could carry out the process well. For this reason, it is very important to carefully design both pre-service and in-service teacher training.
- The problem-solving-based learning and teaching environments should be designed according to the cognitive characteristics of the classroom. Exposing students with high-level problem-solving skills to this approach in a cognitively low-level class will not only adversely affect their ability to solve the related problem, but will also cause their motivation for the lesson to turn negative.
- The methodologies and methods used in qualitative and quantitative research reflect various research strategies and have diverse theoretical, epistemological and ontological concerns. Any strategy will depend on how the researchers gather and analyze their data. To prevent bias in data collection and interpretation, research on problem-solving teaching in science and technology education must be done with prudence.

It is clear that meaningful learning takes place when problem-solving is used effectively and efficiently in science and technology learning and teaching environments. STEM is an effective approach, in which problem solving is an indisputable part of engineering design as described by Kennedy and Odell (Chap. 4 in this book). In addition, students will effectively use their problem-solving skills in solving socio-scientific issues that they will encounter in the future. In future research to be conducted in this field, it would be beneficial to explore different learning environments and student-student and student-teacher collaborations in more depth in order for students to better acquire problem-solving skills.

Summary

This chapter first defined problem-solving and explained its purpose, its use in learning and teaching environments. The chapter discussed the problem-solving methodology and the possible strategies to use problem-solving based on a review of the education literature. In order to provide the learners with a guide, the problem-solving processes, although not necessarily to be followed in a step-by-step fashion, were listed based on several resources describing effective problem solving.

The chapter contributes to the understanding of problem-solving as a teaching and learning method, major European Union projects using problem-solving strategies were investigated. The problem-solving strategies used in these projects were described briefly with references to the project documents or websites.

The chapter ends with the discussions on and suggestions for problem-solving in science and technology education.

Recommended Resources – Books and Journal Articles

The authors recommend the following books and journal articles for further information on assessment and evaluation in science and technology education:

Books:

Wallace, B., Cave, D. & Berry, A. (2009). *Teaching problem-solving and thinking skills through science: Exciting cross-curricular challenges for foundation phase and key stages one and two*. Routledge.

Mackall, D. D. (Ed.) (2004). *Problem solving*. Infobase Publishing.

Sproull, B. (2018). *The problem-solving, problem-prevention, and decision-making guide: organized and systematic roadmaps for managers*. Productivity Press.

Orgoványi-Gajdos, J. (2016). *Teachers' professional development on problem solving: Theory and practice for teachers and teacher educators*. Springer.

Mohanty, N. (2021). *Decision making and problem solving*. Springer.

Journal Articles:

Stewart, J. & Kirk, J. V. (1990). Understanding and problem-solving in classical genetics. *International Journal of Science Education, 12*(5), 575–588.

Larkin, J. H. & Reif, F. (1979). Understanding and teaching problem-solving in physics. *European Journal of Science Education, 1*(2), 191–203.

Gaigher, E., Rogan, J. M. & Braun, M. W. H. (2007). Exploring the development of conceptual understanding through structured problem-solving in Physics. *International Journal of Science Education, 29*(9), 1089–1110.

Gick, M. L. (1986). Problem-solving strategies. *Educational Psychologist, 21*(1–2), 99–120.

Duncker, K. & Lees, L. S. (1945). On problem-solving. *Psychological Monographs, 58*(5), i.

References

Anderson, H. O. (1967). Problem-solving and science teaching. *School Science and Mathematics, 67*(3), 243–251. https://doi.org/10.1111/j.1949-8594.1967.tb15151.x

Ausubel, D. P. (1968). *Educational psychology: A cognitive view*. Holt, Rinehart and Winston.

Binkley, M., Erstad, O., Herman, J., Raizen, S., Ripley, M., Miller-Ricci, M., & Rumble, M. (2012). Defining twentyfirst century skills. In P. Griffin, B. McGaw, & E. Care (Eds.), *Assessment and teaching of 21st century skills* (pp. 17–66). Springer.

Bransford, J. D., & Stein, B. S. (1984). *The IDEAL problem solver: A guide to improving thinking.* W.H. Freeman & Co.

Chi, M. T. H., Feltovich, P. J., & Glaser, R. (1981). Categorization and representation of physics problems by experts and novices. *Cognitive Science, 5,* 121–152.

Chin, C., & Chia, L. G. (2006). Problem-based learning: Using ill-structured problems in biology project work. *Science Education, 90*(1), 44–67.

Egger, A. E., & Carpi, A. (2008). Data analysis and interpretation. *Visionlearning, POS-1,* (1).

Gallagher, S. A., Stepien, W. J., & Rosenthal, H. (1992). The effects of problem-based learning on problem solving. *Gifted Child Quarterly, 36*(4), 195–200.

Gallagher, S. A., Sher, B. T., Stepien, W. J., & Workman, D. (1995). Implementing problem-based learning in science classrooms. *School Science and Mathematics, 95*(3), 136–146.

Garrett, R. M. (1986). Problem-solving in science education. *Studies in Science Education, 13,* 70–95.

Glaser, R. (1992). Expert knowledge and processes of thinking. In D. F. Halpern (Ed.), *Enhancing thinking skills in the sciences and mathematics* (pp. 63–76). Erlbaum.

Greenwald, N. L. (2000). Learning from problems. *The Science Teacher, 67*(4), 28–32.

Hobden, P. (1998). The role of routine problem tasks in science teaching. In B. J. Fraser & K. G. Tobin (Eds.), *International handbook of science education, Vol. 1* (pp. 219–231).

Ioannidou, O., & Erduran, S. (2021). Beyond hypothesis testing. *Science & Education, 30,* 345–364. https://doi.org/10.1007/s11191-020-00185-9

Jonassen, D. H. (1997). Instructional design models for well-structured and ill-structured problem-solving learning outcomes. *Educational Technology Research and Development, 45*(1), 65–94.

Koberg, D., & Bagnall, J. (1981). The design process is a problem-solving journey. In D. Koberg & J. Bagnall (Eds.), *The all new universal Traveler: A soft-systems guide to creativity, problem-solving, and the process of reaching goals* (pp. 16–17). William Kaufmann Inc.

Lawson, M. J. (2003). Problem solving. In J. P. Keeves et al. (Eds.), *International handbook of educational research in the Asia-Pacific region* (*Springer International Handbooks of Education, vol 11*). Springer. https://doi.org/10.1007/978-94-017-3368-7_35

Mahanal, S., Zubaidah, S., Setiawan, D., Maghfiroh, H., & Muhaimin, F. G. (2022). 'Empowering college students' Problem-solving skills through RICOSRE'. *Education Sciences, 12*(3), 196.

McComas, W. F. (1998). The principal elements of the nature of science: Dispelling the myths. In W. F. McComas (Ed.), *The nature of science in science education* (pp. 53–70). Springer.

Milopoulos, G., & Cerri, L. (2020). *Recommendation for future use.* EPINOIA S.A.

Murphy, P., & McCormick, R. (1997). Problem solving in science and technology education. *Research in Science Education, 27*(3), 461–481.

Nezu, A. M. (2004). Problem solving and behavior therapy revisited. *Behavior Therapy, 35*(1), 1–33. https://doi.org/10.1016/s0005-7894(04)80002-9

OECD. (2013). *PISA 2012 assessment and analytical framework: Mathematics, Reading, science, problem solving and financial literacy.* OECD. https://doi.org/10.1787/9789264190511-en

Osborn, A. (1953). *Applied imagination.* Charles Scribner.

Osborne, J., & Dillon, J. (2008). *Science education in Europe: Critical reflections.* Nuffield Foundation.

Pérez, D. G., & Torregrosa, J. M. (1983). A model for problem-solving in accordance with scientific methodology. *European Journal of Science Education, 5*(4), 447–455. https://doi.org/10.1080/0140528830050408

Pizzini, E. L. (1989). A rationale for and the development of a problem-solving model of instruction in science education. *Science Education, 73*(5), 523–534.

Presseisen, B. Z. (1985). *Thinking skills throughout the curriculum: A conceptual design.* Research for Better Schools, Inc.

Sampson, V., Enderle, P., & Grooms, J. (2013). Argumentation in science education. *The Science Teacher, 80*(5), 30.

Simon, H. A. (1973). The structure of ill-structured problems. *Artificial Intelligence, 4*(3–4), 181–201.

Taconis, R. (1995). *Understanding based problem solving*. [Unpuplished PhD thesis],. University of Eindhoven.

Taconis, R., Ferguson-Hessler, M. G. M., & Broekkamp, H. (2001). Teaching science problem solving: An overview of experimental work. *Journal of Research in Science Teaching, 38*(4), 442–468.

von Hippel, E., & von Kroch, G. (2016). Identifying viable "need-solution pairs": Problem solving without problem formulation. *Organization Science, 27*(1), 207–221. https://doi.org/10.1287/orsc.2015.1023

Woods, D. R. (1987). How might I teach problem solving? *New Directions for Teaching and Learning, 30*, 55–71.

Professor Dr. Bulent Çavaş completed his Master and PhD studies in the field of science education at Dokuz Eylul University, Faculty of Education, Science Teacher Training Program in 1998. He did his post-Doc studies at the Middle East Technical University. He has written over 150 national and international publications and 10 books on science and science education. He is Secretary of the CMAS-Science Committee. Currently, his research interests are Responsible Research and Innovation, Open Schooling, Inquiry-Based Science Education and Virtual Reality in Science Education. He is one of the past Presidents of the International Council of Associations for Science Education (ICASE: www.icaseonline.net). He works as an external expert for evaluating European Commission Projects in Brussels, Belgium. Currently, he is working as Professor of Science Education at Dokuz Eylul University (www.deu.edu.tr) in Izmir-Turkey.

Professor Dr. Pınar Çavaş has a Bachelor's degree from Ege University, Physics Department. She continued her Master's studies at Ege University on primary education (major: science education). She completed her PhD degree from Dokuz Eylul University working on elementary teachers' scientific literacy level. She has written more than 100 journal articles, conference papers, books and book chapters. She was involved in some European Union projects, as well as University supported projects. Her fields of interest are primary science education, scientific literacy, and teachers' competences. In addition to her director role at the Children's Education Application and Research Center, she is heading the Department of Primary Education, Faculty of Education, Ege University (www.ege.edu.tr).

Associate Professor Dr. Yasemin Özdem Yılmaz is a science education researcher at the Department of Primary Education at MSKU. She received her Bachelor of Science degree in elementary science education in 2003, and her PhD in Science Education in 2014 at the Middle East Technical University (METU), Ankara, Turkey. During her graduate studies, she was awarded a position by the European Association for Science Education Research (ESERA) to conduct argumentation studies at Bristol University. After gaining her PhD, Dr. Yilmaz studied as a postdoctoral researcher at Great Lakes Science Centre (GLSC), Cleveland, OH for a year (2014–2015). She was a researcher in national and international studies and projects on Argumentation, Inquiry-based learning and Teacher Education in science centers. She is a member of the International Council of Associations for Science Education (ICASE) executive committee, Science Education and Research Association in Turkey (FEAD) executive committee, and Science Teachers Association in Turkey (FENODER). Dr. Yılmaz is still working at Muğla Sıtkı Koçman University (www.mu.edu.tr).

Chapter 19
Creativity and Innovation in Science and Technology Education

Mehmet Aydeniz and Michael Stone

Abstract The STEM education community is in the midst of a paradigm shift. The foundations of traditional instructional context, curriculum, place and pace of learning and methods of learning have been challenged fundamentally. A combination of scholarly efforts and educational initiatives outside of formal education institutions, by entrepreneurs, have disrupted the fundamental assumptions of schooling. These efforts have led to an intentional focus on providing rich opportunities for students to create, collaborate and innovate. In this chapter, we first introduce a discussion related to the concept of creativity. Then, we discuss factors that contribute to individual and group creativity. Next, we introduce one exemplary program from the 'Fab Lab' initiatives. We then elaborate on the design features of these models and discuss how these features empower students to be creative and innovative. Finally, we will discuss the implications for teacher educators, researchers and practitioners, and opportunities for teachers to develop the pedagogical capacity needed for promoting creativity and innovation in their curricula.

Keywords STEM · Creativity · Fab lab · Innovation · Digital skills

Creativity: An Introduction

Creativity is a term that has been frequently used to describe a person, a thought process, or a product. Rhodes (1961) introduced a model of creativity where he subdivided creativity into four Ps: (1) creative Person; (2) creative Process; (3)

M. Aydeniz (✉)
The University of Tennessee, Knoxville, TN, USA
e-mail: maydeniz@utk.edu

M. Stone
Public Education Foundation, Chattanooga, TN, USA
e-mail: mstone@pefchattanooga.org

B. Akpan et al. (eds.), *Contemporary Issues in Science and Technology Education*, Contemporary Trends and Issues in Science Education 56,
https://doi.org/10.1007/978-3-031-24259-5_19

creative Product; and (4) creative Press (conditions). This model suggests that one can find creative features in any of these four Ps. Focusing on the personal characteristics, Harris (1960) defined creativity as *'the ability to produce a number of original ideas when confronted with a problematic situation'* (p.254). Harris assumed that creative engineers: (1) are able to produce more ideas; (2) can change their frame of reference easier and quicker; (3) are more likely to produce uncommon ideas; and (4) are better equipped to visualize in space. While this definition limits creativity to personal characteristics, it suggests that creativity requires unique cognitive attributes and alludes to the domain specificity of creativity. Increasing student creativity has been a focus of K-12 educators and, more recently, educators have engaged in curriculum development in and out of school contexts to promote student creativity (Burke, 2014).

Creativity is an important term and skill across many industries and educational settings, yet K-16 students continue to have limited opportunities to develop this critical skill in formal academic settings. Two of the factors that have limited the teaching of creativity have been lack of resources/tools and time to engage in and finish creative experiences in K-12 settings. In this chapter, we focus on conditions that nurture acquisition of personal creativity, and spaces that allow personal creativity to thrive, through a real-world example. We first discuss factors that impact creativity. Next, we introduce digital tools that enable personal or group creativity. Then, we discuss skills that the students will need in the twenty-first century economy to engage in creative activity, followed by the processes that facilitate creative problem-solving. Finally, we make several recommendations for teacher educators and school administrators.

Factors Impacting Creativity

A review of relevant literature suggests that several factors can impact individual creativity, including context, processes, tools and personal attributes. While we do not aim to discuss the full details of these factors, we will provide an overview of each to guide our readers as an introduction for further exploration.

Context

Contexts that promote individual creativity are those that present disorderly situations and problems that require creative thinking, coupled with access to tools and resources that facilitate creative problem-solving (or creative making). Unfortunately, traditional school settings do not have these characteristics that enable, facilitate, or nurture student creativity. Such contexts include maker spaces or Fab Labs, STEM competitions and internships, among others. One characteristic of these contexts is that they allow students to tinker with their ideas, provide psychological safety for testing out-of-the-box ideas, and encourage authentic collaborative inquiry.

Curricular Focus

The second factor that can contribute to students' acquisition of creative problem-solving skills is the nature of the curriculum to which we expose our students. We know that most traditional curricula fail to afford rich opportunities for students to develop habits and skills for engaging in creative thought, creative problem-solving, and creative making. Firstly, traditional learning environments, by design, prioritize and reward acquisition of content knowledge over skill development. While some schools attempt to elevate skill development, they often default back to overemphasizing content mastery because of the extreme systemic focus placed on content standards and state testing. One thing that we need to keep in mind is that skills development is not a linear, prescriptive experience. Students develop skills through repeated experience, through failures, through collaboration with peers. So, the skills development process is messy and cyclic rather than linear. Secondly, traditional learning environments restrict creative making, as student experiences are bound by rigid school schedules, prescriptive curriculum-pacing guides, and measurements of student success that are solely content-focused. Thirdly, teachers' dispositions, knowledge and skills play a critical role. Most teachers are the products of a system that has emphasized, taught, assessed and rewarded content knowledge over skill development. Consequently, without extensive and explicit professional development, teachers resort to the way in which they learned STEM subjects when teaching STEM concepts.

The sole focus on content acquisition restricts opportunities for students to engage in divergent thinking, which has been associated with creative thought. We also know from the OECD report (OECD, 2014) that most countries' high school students underperform in creative problem-solving, which reflects the focus of curricula and the methods of teaching that fail to provide rich opportunities for students to engage in creative thinking and creative problem-solving. When the curriculum and instruction focus primarily on students' acquisition of scientific facts, and teachers hold only minimal subject matter knowledge, it becomes rare, if not impossible, for students to develop habits of minds, dispositions and skills necessary for creative thinking, creative problem-solving and creative making. Despite these problems, recent developments in STEM education have created contexts and tools for students to engage in creative thinking, creative problem-solving and creative making. We elaborate on these developments in the next section. However, we first introduce personal and contextual attributes that are associated with creative problem-solving.

Personal Characteristics

Personal characteristics that promote creativity include, but are not limited to: open-mindedness, curiosity, problem-solving skills, persistence, comfort with ambiguity, and metacognition. We must note, however, that these personal characteristics are the outcomes of the experiences the individuals have had in contexts that enable, facilitate and nurture creative thought and creative problem-solving. These characteristics are developed through consistent engagement in rich experiences that call for creative thinking, creative problem-solving and creative doing. Access to epistemic, mentoring, academic peer groups, and physical resources, makes a difference in how one thinks in academic and non-academic environments. Readers should keep this perspective in mind as they make sense of what we present in the following sections.

- *The first* of these personal characteristics is *curiosity*. **Curiosity** refers to the level of discomfort with a gap in knowledge or the joy of and passion for exploring the unknown. *Curiosity* feeds creativity, because it encourages and ensures: (1) sustained inquiry mindset and behavior; (2) divergent thinking, which can lead to development of alternative approaches to problem-solving; (3) pursuit and development of alternative explanations for the observations; and (4) risk-taking. All of these are associated with creative thought, creative problem-solving and creative making.
- *The second* personal characteristic related to creativity is *metacognition.* *Metacognition* empowers creative people to see aspects of their cognition that facilitate, help and encourage problem-solving. It allows them to see gaps in their problem-solving journey, and gives them the opportunity to reflect on their knowledge and methodology. Therefore, any educational endeavor aiming to promote student creativity should focus on cultivating metacognition, as it is a critical component of learning through rich experience.
- *The third* personal characteristics is *open-mindedness. Open-mindedness* refers to the human attribute that allows one to be receptive to a variety of ideas, methods and arguments, and a willingness to consider relevance of alternative strategies to the problem in hand. *Open-mindedness* is a prerequisite to creative thinking, creative problem-solving, and creative making because it allows one to see multiple factors that may contribute to a problem, to consider divergent pathways that could inform a novel solution. Open-mindedness encourages acceptance of failure early on and a willingness to try new and alternative problem-solving methods. It also encourages use of alternative resources to achieve a creative goal, or to propose creative solutions to a complex problem.
- *The fourth* personal quality of creative people is grit or *persistence. Grit* refers to the ability of an individual to endure challenges and persist over time on a journey towards accomplishing important goals:

  > 'You have to be burning with an idea, or a problem, or a wrong that you want to right. If you're not passionate enough from the start, you'll never stick it out' (Steve Jobs).

People develop creative solutions and products partly because they do not give up easily. **Grit or persistence** is a personal quality that discourages people from giving up on problem-solving in the face of failures or adversaries. Instead of giving up, creative people model persistence as they work through multiple iterations, choosing to use failed attempts to inform future strategies, rather than giving up after initial methods do not yield the desired result.

Tools for Creativity

Digital learning has opened immense opportunities for teachers to design activities for promoting student creativity. Students also have access to a set of digital tools that they can use to engage in creative activity in the absence of a pedagogical guide such as the teacher. Firstly, digitized information/content can be accessed asynchronously and independently. This alleviates the need for learning within a rigid academic schedule and it frees the teacher to focus on skill development instead of disseminating content. Secondly, digital communication tools make collaborative problem-solving possible. They give immediate access to community resources and democratize where and how students learn – simultaneously providing access to global experts and local advocates. Students can interact with each other, give feedback, ask questions, and have immediate access to epistemic resources to make connections, address a knowledge gap that they may have, and access a diverse and robust set of relevant resources. One of the best examples of how digital tools can promote student creativity is the *Scratch* community. Digital tools allow for community-building, sharing of community resources, collective problem-solving and epistemic affordance. Collectively, these features of digital technologies make creativity possible. However, this possibility alone is not sufficient; the experiences should be scaffolded for creative thought, creative problem-solving, and creative products. Teachers should have the disposition, domain knowledge and pedagogical skills to design and facilitate learning activities for students to develop creative thought, creative problem-solving, and/or creative making. As teacher educators and administrators, we should help teachers develop such dispositions, domain knowledge and pedagogical skills so that they can effectively guide their students' skills acquisition. This can take place through ongoing professional development and community-building in disciplinary and interdisciplinary contexts.

Digital Skills for Creativity

While digital tools help students to collaborate more effectively across contexts, access important resources and more effectively learn about the abstract concepts, the role of technology should not be limited to accessing and sharing knowledge between the learners. Schools or educational entities should create contexts, space and learning activities that will allow students to develop digital skills for engaging in creative problem-solving and creative making. Paired with progressive pedagogical strategies, these spaces empower students to realize the full potential of digital skills. The first of these skills is computational thinking.

- **Computational thinking skills**

Computational thinking refers to the type of skills that involve the use of computational tools and computing power to solve real-world problems or to design services, experiences and products (Wing, 2006). We now live in a digital world. The breadth of our experiences is tailored for technologically rich spaces. The next generation already lives in the digital world through video games and VR. With the growing popularity of metaverse, social aspects of our lives will increasingly migrate deeper into digital space. As our experiences move to digital spaces, the new economy becomes the economy of makers. More specifically, it becomes the economy of making by programming or coding using digital tools. In order for us to prepare our students for this type of economy, we need to integrate computational thinking and coding skills across the curriculum, rather than isolating it to traditional STEM fields.

- **Data science skills**

The second skill is data science. In addition to computational thinking skills, we need to teach our students data science, data engineering and data management skills. These skills collectively enable students to practice with creative design, creative problem-solving, creative modeling, and creative making. These skills are the fuel of the new economy; therefore, any creative design will depend on computational and data science skills (Fig. 19.1).

- **Collaboration skills**

Another important skill for creative problem-solving is collaboration. Collaboration is critical across disciplines, industries, borders, contexts and skills. The new generation of employees must develop the ability to work collaboratively. Collaboration requires being open-minded and having excellent communication skills. Engaging students in collaborative learning and collaborative problem-solving early on not only helps students develop knowledge and skills, but also cultivates positive dispositions towards communal growth. Collaboration increases students' metacognition, as they monitor and evaluate their contribution to the project, and how those contributions serve or do not serve the accomplishment of the goal. They learn to integrate knowledge and skills across different domains, gain exposure to different perspectives, and learn to use evidence and data to share, challenge and defend ideas presented to the group.

- **Design thinking skills**

The fourth skill associated with creativity is design thinking. Design thinking draws on data, human imagination, and systematic reasoning to explore creative answers to complex problems. Design thinkers imagine creative possibilities

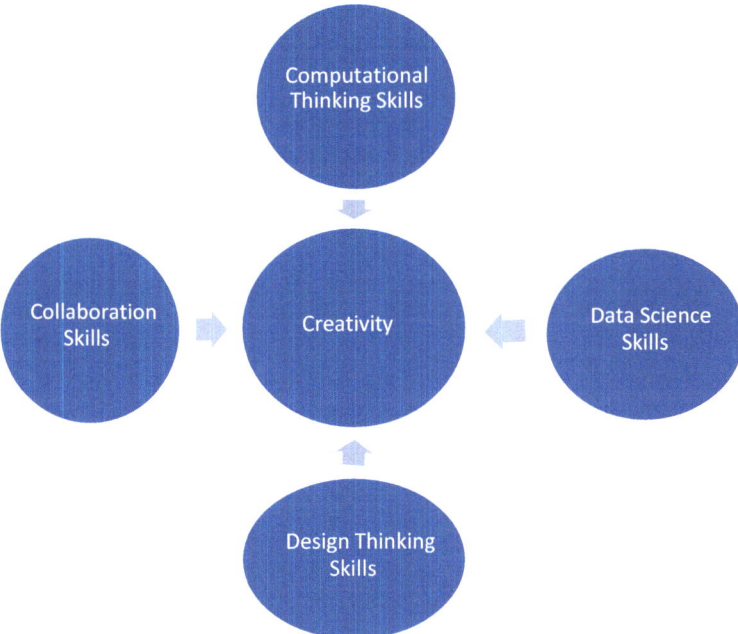

Fig. 19.1 Digital skills

informed by data and contextual expertise. There are multiple models and definitions of design-based thinking, but the core aspects of design thinking are: empathizing (understanding the problem through lived experiences), defining (which corresponds to defining stakeholders, challenges, roles and opportunities), and ideation. The next stage is prototype development. At this stage, the individual is expected to develop a prototype for testing the specifications of the target design product or service. The final stage of design thinking involves testing the product for its effectiveness, particularly through a lens of empathy for the end user. After the initial testing, the process is repeated to optimize the effectiveness and efficiency of the final design.

In a data-driven world, where computational power and tools are abundant and human experience has moved to a digital plane, these four fundamental skills should be at the core of any curriculum that aims to promote student creativity.

Processes That Facilitate Creative Problem-Solving

There are several processes that facilitate creative thought, creative problem-solving and creative making. These include collaborative inquiry, ***collaborative problem-solving***, opportunity for ***reflection on experience,*** and ***receiving feedback or*** *criticism*. School curriculum and instructional strategies should facilitate student engagement with processes that promote creative problem-solving skills through authentic collaborative inquiry.

The First Process That We Believe Facilitates Such Opportunity Is *Collaborative Inquiry* When students engage in collaborative inquiry, they build on each other's contributions to advance the arguments or to improve their design or the quality of their arguments of products. Similarly, different members of the group can make different observations and highlight issues that otherwise may not surface. Each member can build on their observations, unique experiences and prior knowledge to ask different questions that can result in new knowledge. This new knowledge can be integrated to inform the design, product, or solution. The collaborative inquiry experiences can enrich students' domain knowledge repertoire, expose them to alternative explanations, and raise diverse questions, which, collectively, can influence the quality and effectiveness of the products, arguments and models.

The Second process That Helps Facilitate Creative Thought Is Reflection Reflection time is typically limited in traditional classroom settings, as the bell schedule does not allow for reflection over an extended period. Digital learning environments overcome this limitation, as the experience of learning is not limited to a typical class schedule. Moreover, the triggers of self-reflection in the classroom are limited to the teacher and classmates. However, in social digital communities, students have access to the resources and questions of a larger, global community. This provides a unique opportunity to reflect in a more diverse and more informed context.

The Third Process That Facilitates Creative Thinking Is Argumentation Argumentation provokes creative thought and creative problem-solving. Argumentation allows the learner to integrate knowledge across different domains to develop an articulate argument, encourages logical conclusions when presenting one's arguments, and clearly articulates the argument's rationale in an effort to demonstrate transparency and encourage critical discourse. Argumentation and the criticism received from peers can help the learner to identify gaps in their knowledge or models, deficiencies in their reasoning, and the quality and relevance of their evidence. This critical inquiry into one's evidence, model, reasoning, or argument is generative and cultivates creativity as students engage in authentic experiences.

After indicating the contexts, personal attributes and processes that contribute to creative problem-solving, we will now introduce the *Maker Movement,* which has allowed student creativity to thrive. After presenting a brief discussion of what the Maker Movement is, we provide a real example of what this type of education makes possible for students (Fig. 19.2).

Fig. 19.2 Processes

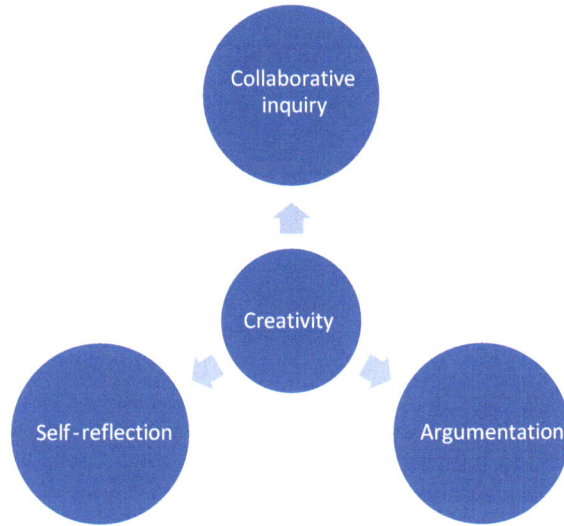

Maker Movement

Maker education is a movement that aims to empower students to develop robust skills as they design and create tangible products using imagination, creativity and technology. Maker spaces provide access to digital tools, materials and software necessary for students to develop functional solutions to real-world problems. One key aspect of maker spaces is that they encourage community building and collaborative learning (Burke, 2014). The Maker Movement has permeated formal and informal learning environments. Spaces have been embedded in museums, science centers, libraries, schools and community centers. Schools have been working to repurpose some of their classrooms to accommodate the rich learning that can occur in a maker space. Similarly, schools have been more purposeful in recruiting teachers who can help run the maker spaces and provide intentional guidance to their students to make the learning experiences meaningful and powerful.

- **A Makerspace example**

What content, skills and habits should all students master in school? It is clear that traditional content is important for student development, but it is also critical that students develop strong habits and skills. The team at the Public Education Foundation (PEF) in Chattanooga, TN, asked this question of more than 300 leaders from business, industry and higher education. Nearly all responses indicated skills and habits – so-called *STEM Essential Skills*. In particular, two primary categories emerged from the responses. Leaders are clamoring for students who have strong *interpersonal* skills (i.e., collaboration and communication) and strong *learning* skills (i.e., critical thinking, adaptability and creativity). Unfortunately, because content mastery is simpler to quantify and assess, over the last few decades the education pendulum has become stuck on the wall of content. Certainly, content areas such as mathematics, history, language and science are important. However, informal polling suggests that these STEM Essential Skills are at least as important as content mastery. This observation pushed the team at PEF to begin redesigning learning experiences that elevate opportunities for students to develop STEM Essential Skills alongside traditional content mastery.

Dr. Tony Donen, founding principal of STEM School Chattanooga (STEM Chatt), worked in partnership with Michael Stone, VP of Innovative Learning at PEF, to design these innovative learning experiences. STEM Chatt was founded as a platform school intended to identify, develop, incubate and deploy innovative learning strategies for public high school students. Based on the polling mentioned above, Dr. Donen led the original faculty to identify core tenets that became the hallmark of the school. The team chose critical thinking, creativity, and collaboration – STEM Essential Skills – as their core values. To ensure that students could begin developing these competencies alongside traditional content, the team decided to use multi-disciplinary, project-based learning (PBL) throughout the school. Additionally, they developed a faded scaffolding approach that strategically releases ownership and autonomy to the students in every facet of their high school experience. Two years into opening the school, one grade level per year, the leadership team realized that students were progressing in their development of essential skills, but their PBL presentations were void of functional solutions. The students would present rough analog models and nicely designed slide-decks, but they never had an opportunity to engage in design thinking where they could test and analyze their proposed solutions to real problems. They were not being afforded the opportunity to work through an iterative design process and glean understanding from the incidental learning moments that naturally occur in these strategic experiences. In researching opportunities associated with the Maker Movement, the team stumbled onto the Fab Foundation and the Fab Lab model that had been developed at MIT.

After some strategic design sessions, the team developed a plan to embed a Fab Lab into the high school using what is now considered an 'open lab' model. In this model, the lab serves as a room containing rapid prototyping tools, where students develop functional solutions to authentic problems as an integral part of their PBL experience. To embed this model, Dr. Donen made it clear that the goal of the lab was not to explicitly teach discrete technical skill sets (like 3D design or physical computing). Instead, the aim was to provide opportunities for students to develop STEM Essential Skills. Using the advanced technology in the lab (labs are fitted with 3D printers, laser cutters, vinyl cutters, CNC machines and physical computing components), the aim shifted from content mastery to empowering students to leverage the resources around them to quickly learn new skills and solve real problems. Students have full access to the lab as they engage in their multi-disciplinary PBL units. The teachers work to facilitate a 'just-in-time' learning environment where students acquire technical knowledge and skills as they are needed in their design process. To this end, teachers coach student teams through PBL product development, pointing them to resources instead of serving as the sole or primary access point for knowledge and information. Additionally, the Fab Lab teacher focuses all assessment solely on student mastery of STEM Essential Skills. Rather than assessing product quality or functionality, technical knowledge or content mastery (all discrete components whose assessment often discourages creative, innovative solution attempts), the Fab Lab teacher solely focuses on coaching and assessing student mastery of the targeted essential skills.

This explicit focus on essential skill development represents a subtle but powerful shift in student development. For too long, schools have solely focused on *what* students know, when the important question is actually *what can they learn and do*? Can they ask thoughtful questions, access relevant information, interpret it, analyze it, and then do something with it? Schools shouldn't be simply measuring what information students can recall. They should be measuring if students can leverage essential skills to ask relevant questions and then create something meaningful with the information that they discover.

To bring this model to scale in more traditional schools, the team had to recognize a critical and necessary shift. Much of modern schooling is still designed around relics of the **factory model** of education tailored for the industrial era. In this model, teachers serve as *content* experts who share knowledge with students in order to prepare them for specific, predictable roles in a relatively slow-moving world.

However, the modern world is dynamic. Today, students don't need teachers to serve as gatekeepers to information. They can't afford to sit through a static learning model. They need an adaptive model, where teachers serve as *learning* experts who empower students to thrive as learners and doers. Students need opportunities to imagine creative solutions to complex problems, and then bring those visions to life. Central to this model is a critical shift in the role of the teacher. Students need to learn *with* teachers, not *from* them. They need teachers to model how to use essential skills to thrive as agile learners who quickly learn new skills, apply innovative approaches to novel problems, and succeed in collaborative environments.

Unfortunately, many classrooms still reflect the didactic, prescriptive models designed for a world that passed us by decades ago. Mostly, if not entirely, content-focused, school systems and entire cottage industries work to prepare students to regurgitate information in a futile attempt to beat computers at fact recall and calculations. It is as if we have forgotten about the value of creativity and critical thinking entirely. Sadly, this is not a novel observation. Nearly 16 years ago, the late Sir Ken Robinson gave arguably the most famous TED Talk of all time as he made a moving case for embracing creativity and rethinking the goal of modern schooling. Around the same time, Tony Wagner began pushing for schools to embrace what he calls the *7 Survival Skills for the twenty-first Century:* competencies like critical thinking, agility and communication, which we know are important for student development, but which we have yet to effectively integrate into the learning experiences of our formal educational institutions.

In 2014, in Mr. Stone's first semester at STEM Chatt where he was hired to open the Fab Lab as the first teacher in the space, he met a student named Emma *(name changed to protect privacy)*. She was a 16-year-old high school junior who didn't identify as particularly tech savvy or mechanically inclined, but she had an experience in the school's Fab Lab that dramatically impacted her future. When presented with an opportunity to contribute ideas to a local art installation, Emma imagined what it might look like to build an ice castle for the company's holiday window display. In just 6 weeks, Emma learned enough design software to create a table-top model by laser-cutting a few pieces of acrylic and fitting them together to make a facade of a 'castle'. She then joined with other students in the business partner's boardroom to present her team's design. Her team was the last to make a presentation, and you could feel the excitement behind their nervousness as they began to share the model. A few minutes into the presentation, the Chief Executive interrupted to ask Emma and her team if they could 'actually build this to scale'. Emma didn't miss a beat. She eagerly responded 'YES' (despite lacking the discrete technical skills necessary to build the display). Over the next 5 weeks, Emma **collaborated** with a few engineers, **accessed** online resources to learn enough computer-aided design (CAD) to use the Fab Lab's computer-controlled router, **imagined** and **experimented** with how light interacts with different materials, and grew to **lead** her team to build a 12-feet tall by 10-feet deep by 30-feet long acrylic ice castle (see Fig. 19.3). By blending the essential skills that she had developed at STEM Chatt with access to the advanced technology in the school's Fab Lab, she was able to move from an idea in her head to a tangible, stunning solution.

When the design was revealed at a press event hosted by the company, Emma became an instant hit. The media rushed to get photos and interviews as her friends and family watched from the sidelines, beaming with pride. The moment was truly fantastic, but perhaps a bit fleeting. A week later, Emma joined her classmates as a new business partner pitched the next challenge. However, this project was a little less artistic. A local caving tour company explained that the students would design and create potential solutions to mitigate the spread and impact of white-nose

Fig. 19.3 Student product

syndrome (a highly contagious and deadly fungus that was beginning to afflict bat populations in the region). Despite the clear disconnect from the artistic ice castle model, Emma was ready to jump in! When she returned to school, she went straight to the Fab Lab, where she transferred the same skills that she had used in the previous project, **collaboration** and **leadership,** to rally her team to accomplish the task, **accessing** and **analyzing** new content to learn details about how bat colony behaviors aid the spread of the fungus, and engaging in **critical thinking** and **agility** to ask insightful questions and **imagine** potential solutions.

While the media celebrated the artistic value of the ice castle, classroom experiences that celebrated *process* over product and *essential skill development* over memorization equipped Emma with the confidence and capacity to create solutions to authentic problems – whether building a beautiful ice castle or mitigating the spread of a bat fungus.

Interestingly, this focus on essential skills doesn't have to come at the cost of content mastery. The two are not mutually exclusive. Emma went on to earn a Bachelor's degree and is now thriving in a full-time role at a national avionics company. Additionally, the opportunity that she had at STEM Chatt has grown from a pilot program at one extraordinary public school to a burgeoning movement, scaling to 30 K-12 public schools in Hamilton County, TN, and now growing to at least 19 additional labs across the country. Today, more than 45,000 students in public schools, spread across diverse communities throughout the United States, are using Fab Labs to cultivate essential skills through authentic learning experiences.

Pedagogical Design Features of the Learning Contexts That Promote Creativity

There are several design features of these learning environments. We discuss each of these design features below.

Creative Challenge

Students must be presented with a challenge or a problem that requires creative thinking, creative problem-solving, or creative making. If the curriculum engages students only in routine learning tasks that require memorization, students will not learn to engage in creative problem-solving. While traditional classrooms often cannot afford such opportunities, teachers can leverage constructivist learning strategies to overcome the challenges presented by traditional curricular goals and structures that prohibit teaching of creativity.

Project-Based Learning

Learning often occurs in a project-based format. Project-based learning (PBL) is an important pedagogical design feature of these environments. Firstly, PBL engages students in inquiry-based learning. Students engage in open-ended problems that facilitate rich learning experiences and extend well beyond the traditional class period. Secondly, PBL promotes student curiosity, as the problems are intentionally designed without simple, clear solutions. Students must think deeply about the problem, methods, data and efficacy of models proposed as solutions. Thirdly, the process of learning is collaborative. Students engage in brainstorming during the planning stage, and in argumentation over data, methods and models developed as a solution. They develop alternative explanations and challenge one another about the process and products of their inquiry. Creativity emerges as a natural bi-product of these processes.

Psychological Safety

Additionally, psychological safety is a key attribute of environments that promote creativity. When the learning culture provides psychological safety, students can think freely, express unorthodox ideas with their friends and teachers, and avoid the fear of making mistakes. In such learning environments, even the wildest ideas are considered with an open mind, and are discussed without cognitive bias. Collectively, these experiences and perceptions encourage divergent thinking and lead to student creativity.

Conclusion and Implications

In this chapter, we focused on factors that encourage student creativity. We will now discuss the implications of this understanding for teacher education, research and practitioners.

Teacher educators need to understand that learning is no longer restricted to the brick walls of classrooms. Learning is taking place everywhere, experienced by everyone, and taught by everyone. We must prepare teacher trainees for this reality. They should be exposed to innovative learning contexts, experience of digital tools, and communities that promote student creativity through both structured and unstructured learning tasks.

School administrators should understand that limiting learning to teacher lectures and designing learning goals around student test scores will only limit opportunities for students to learn and advance in their professional careers. The narrow focus on test scores should be mitigated by opportunities for students to develop creativity and other essential skills, alongside content mastery.

School leaders should be aiming to promote students' skill development. They should communicate the expectations to their teachers so that they can cross borders, moving from test-focused instruction to a skills-focused model.

School administrators should support professional learning opportunities for their teachers, through which they will acquire knowledge and skills related to designing rigorous learning environments, rich in content *and* interactions, to support student creativity through digital and physical tinkering, and through project-based and design-based activities.

Researchers should develop new assessments consistent with the evolving goals of STEM education: the transition from content understanding to skills development.

Summary

This chapter primarily covers the meaning of creativity, factors impacting creativity; tools for promoting creativity in STEM learning environments, processes that facilitate creative problem-solving, and learning contexts that promote creativity.

References

Burke, J. J. (2014). *Makerspaces: a practical guide for librarians (Vol. 8)*. Rowman & Littlefield.
Harris, D. (1960). The development and validation of a test of creativity in engineering. *Journal of Applied Psychology, 44*(4), 254–257.
OECD. (2014). *PISA 2012 results: Creative problem solving: Students' skills in tackling real-life problems (volume V)*. OECD Publishing.
Rhodes, M. (1961). An analysis of creativity. *The Phi Delta Kappan, 42*(7), 305–310.
Wing, J. M. (2006). Computational thinking. *Communications of the ACM, 49*, 33–35. https://doi.org/10.1145/1118178.1118215

Further Readings

Bemiss, A. (2019). *Inspiring innovation and creativity in young learners transforming STEAM education for pre-K-grade 3*. Routledge. https://doi.org/10.4324/9781003235811
Dorst, K., & Cross, N. (2001). Creativity in the design process: Co-evolution of problem–solution. *Design Studies, 22*(5), 425–437.
Stretch, E., & Roehrig, G. H. (2021). Framing failure: Leveraging uncertainty to launch creativity in STEM education. *International Journal of Learning and Teaching, 7*(2), 123–132.

Lin, Y. (2011). Fostering creativity through education--a conceptual framework of creative pedagogy. *Creative Education, 2*(3), 149–155.

Dr. Mehmet Aydeniz is professor of STEM Education at The University of Tennessee, Knoxville. He conducts research related to student learning and teacher practice in STEM education. He is mostly interested in helping teachers to develop self-efficacy in STEM pedagogies related to scientific practices in K-12 classrooms. Dr. Aydeniz is the Editor of the *Journal of Research in STEM Education.*

Mr. Michael Stone is the Vice President of Innovative Learning at the Public Education Foundation in Southeast Tennessee. He joined in 2016, after concluding an appointment as an Albert Einstein Educator Fellow at the National Science Foundation. He spent the first 10 years of his career as a high school mathematics and computer science teacher. In his current role, he has led the development of the largest school-based Fab Lab network in the world. This work has garnered international attention, allowing him to present in numerous countries and venues, including presentations at the White House and a US Senate Briefing. He has published three books, including *Let Me Try It: Enhancing Maker Education through Digital Fabrication.*

Chapter 20
Collaboration and Communication in Science and Technology Education

Michael R. L. Odell, Kelly Dyer, and Mitchell D. Klett

Abstract Communication and collaboration are twenty-first century skills that many consider to be essential for students in science and technology education. Communication and collaboration are interdependent skills that can be used to promote deeper learning in science through active pedagogies such as project-based and problem-based instruction. Providing students with the opportunity to communicate science through the development of scientific posters, manuscripts for publication and multimedia development can enhance science content and develop skills that are useful beyond the science classroom, and which are used in personal and professional life.

Keywords Communication · Collaboration · Twenty-first century skills · Project-based learning · Problem-based learning · Science communication · Active learning · Collaborative groups

Introduction

Communication in science and technology is considered an essential skill for students and STEM professionals alike. What is communication in science and technology? *'Science communication has many definitions and not all researchers and practitioners agree on its goals and boundaries'* (Gascoigne & Schiele, 1995). As a result, there can be confusion regarding definitions that are used in the research literature.

M. R. L. Odell (✉) · K. Dyer
The University of Texas at Tyler, Tyler, TX, USA
e-mail: modell@uttyler.edu; kdyer@uttyler.edu

M. D. Klett
Northern Michigan University, Marquette, MI, USA
e-mail: mklett@nmu.edu

© The Author(s), under exclusive license to Springer Nature Switzerland AG 2023
B. Akpan et al. (eds.), *Contemporary Issues in Science and Technology Education*, Contemporary Trends and Issues in Science Education 56, https://doi.org/10.1007/978-3-031-24259-5_20

283

For the purposes of this chapter, science communication will focus primarily on communicating science to the public. Excellent public engagement with science builds on a foundation of clear and concise communication. What are the skills necessary for effective engagement of the public when presenting scientific topics? These skills complement the science and technology teaching strategies presented in Chap. 15 of this book.

Aurbach et al. (2019) developed a framework of the foundational skills necessary for science communication. They recommended that foundational communication skills can be separated into distinct categories including identifying appropriate communication or engagement goals and objectives; adapting to a communication landscape and audience; messaging; narrative; language; visual design; non-verbal communication; writing style; and providing a space for dialogue.

The American Association for the Advancement of Science (AAAS) provides support to scientists and journalists to improve science communication. When communicating science, AAAS recommends that communicators find messages that are 'short and simple'. They also recommend that those communicating science present no more than three ideas, so that these ideas are more memorable to audiences. We take this idea a step further, to go beyond science communication to the general public, but also to students enrolled in primary, secondary and tertiary education.

Communicating science effectively can enable meaning engagement by non-scientists. Those of us in the science or science education professions regularly present science to non-scientists. Learning to effectively communicate science to students in the tertiary education system, especially to students who will become scientists or elementary and secondary teachers, is important. In fact, an advantageous quality of foundational communication skills is that they can be used regardless of audience and should be incorporated into education training (Brownell et al., 2013).

In addition, the authors of this chapter believe that preparing students for the communication of science using multiple platforms (writing, video, presentations, etc.) should extend into the instruction of primary and secondary science students. Science communication is becoming increasingly recognized as a core skill or competency in the science curriculum. Developing K-12 students' science communication skills aligns well with twenty-first century skill development to improve science and technology learning outcomes. Effective communication also facilitates better collaboration, another skill that has been seen as essential in the endeavor of science. The development of the twenty-first century skills framework highlights the importance of communication and collaboration for education and, ultimately, in the workforce or public setting.

Historical and Theoretical Background

The development of a framework of twenty-first century skills is well documented in the science and technology education literature. The most commonly referenced framework was developed by the PS 21 Network. The P21 Frameworks for twenty-first Century Learning were developed by gathering input from a large group of stakeholders, including teachers, education, business and industry leaders to define the skills and knowledge that students need to be successful in school and in life. This includes workforce skills. The Frameworks are illustrated in Fig. 20.1.

Of particular interest are the Learning and Innovation Skills. These include *Critical thinking and problem-solving; Creativity and innovation; and Communication and collaboration.*

This chapter focuses on the third Learning and Innovation Skill, *communication and collaboration.* The PS 21 Framework for twenty-first Century Learning Definitions document (2015) provides clear definitions of communication and collaboration skills to prepare students to communicate clearly. Individual skill definitions include: Articulate thoughts and ideas effectively using oral, written and non-verbal communication skills in a variety of forms and contexts; Listen effectively to decipher meaning, including knowledge, values, attitudes and intentions; Use communication for a range of purposes; Utilize multiple media and technologies, and know how to judge their effectiveness; Communicate effectively in diverse environments; Collaborate with others; Demonstrate ability to work effectively and respectfully with diverse teams; Exercise flexibility and willingness to be helpful in making necessary compromises to accomplish a common goal; and Assume shared

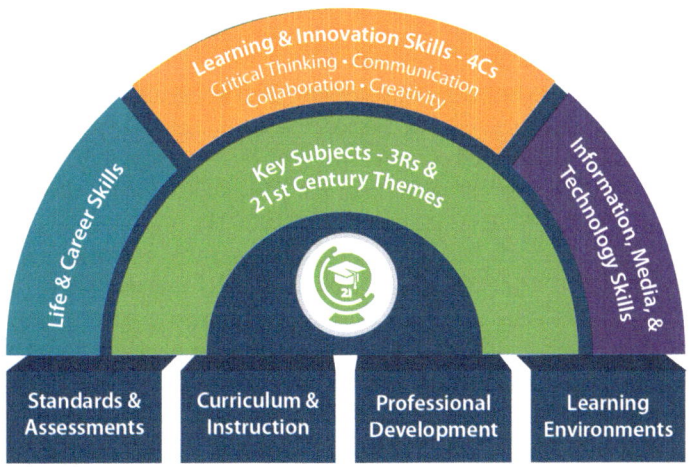

Source: http://www.battelleforkids.org/networks/p21/frameworks-resources

Fig. 20.1 Twenty-first century learning frameworks. Source http://www.battelleforkids.org/networks/p21/frameworks-resources

responsibility for collaborative work, and value the individual contributions made by each team member.

Communication and Collaboration in the Context of Twenty-First Century Skills

Put simply, learning science content is no longer sufficient in the current global context in which our students will work. Communication and collaboration are two of the twenty-first century skills that are required to participate fully in an increasingly complex science and technology workforce. The teaching of science and technology at the primary and secondary levels must adopt pedagogies that support the incorporation of communication and collaboration skills in the classroom.

Communication and Collaboration in the Science and Technology Classroom

Communication and collaboration are two skills that every science and technology educator utilize every day in the classroom and in their personal life. As noted earlier in this chapter, definitions of collaboration and communication are not well defined and are often used interchangeably in the research literature. That said, there are pedagogies that support the incorporation of twenty-first century skills, including communication and collaboration. Project- and problem-based learning (PBL) are considered foundational pedagogies to prepare students in twenty-first century skills, including communication and collaboration (Barell, 2010). PBL can enhance students' communication and collaboration skills in STEM. Research suggests that PBL can develop students' ability to share and understand ideas, as well as present those ideas and be more receptive to perspectives different from their own (Owens & Hite, 2022). In a randomized control study (RCT), it was found that students who participated in a PBL course out-performed non-PBL students on the US Government and the Environmental AP Exam (Saavedra et al., 2021).

There are school models that utilize PBL as the primary instructional approach in the United States. The New Tech Network (NTN) partners with almost 200 schools to implement PBL and provides schools with rubrics in the areas of communication and collaboration. The rubrics focus on three areas: (1) Interpersonal communication: focuses on the listening and speaking skills exhibited by individual students in a wide variety of informal conversations (e.g., student and teacher, student and student and expert); (2) Presentation: focuses on the elements of a strong and complete presentation; and (3) Delivery: focuses on the individual aspects of a presentation.

The rubrics are designed to provide students with guidance for creating and communicating their work, and to provide teachers with evidence for providing grades. There is also a separate written communication rubric to support teachers and students.

NTN has also developed rubrics for collaboration to assist students to be productive members of teams with the goal of a commitment to shared success, leadership and initiative (New Tech Network, 2016a, b, c). The rubric has two elements: an individual collaboration rubric and a team or group checklist. The individual collaboration rubric focuses on specific aspects, skills and behaviors of individual collaboration. These include interpersonal communication, which overlaps with the communication rubric, commitment to shared outcomes, and team leadership (See Fig. 20.2).

New Tech schools are highly effective school models based on a number of research studies, but they are not the only school model utilizing PBL that fosters the development of twenty-first century skills including communication and collaboration. In the US state of Texas, a 4-year longitudinal study found that T-STEM Academies, schools focused on Science, Technology, Engineering and Mathematics (STEM), had a statistically significant impact for ninth graders in standardized math and science assessments compared to peers in matched schools. Two of the T-STEM

NTN Collaboration Rubric, Grade 12

The ability to be a productive member of diverse teams through strong interpersonal communication, a commitment to shared success, leadership, and initiative.

	EMERGING	E/D	DEVELOPING	D/P	PROFICIENT	P/A	ADVANCE
INTERPERSONAL COMMUNICATION	• Distracts conversations by expresses ideas that are off topic, undeveloped, or based on limited understanding of the topic • Shows little interest in the ideas of others • Asks questions that are irrelevant or distracting • At times, addresses others with disrespectful language or tone. • Monopolizes "air time" or frequently interrupt other speakers		• Sometimes is awkward or has difficulty expressing ideas, but conversations are relevant to the topic and based on facts or evidence. • Listens with partial interest in the speaker's message providing sporadic verbal/ nonverbal feedback to indicates some understanding or agreement • Asks general questions to clarify understanding of speaker's point of view • Usually address others with respect, with minor lapses • Shares "air time" by allowing others to speak		• Contributes to productive conversations by clearly expressing well-developed ideas that are relevant and supported with evidence or sound reasoning • Listens with interest to the ideas of others providing verbal or nonverbal feedback to signal understanding or agreement • Acknowledges and helps clarify the ideas of others by asking probing questions. • Responds to different ideas or opinions with diplomacy • Addresses others with respect and sensitivity to cultural or language background • Works to resolve conflict through productive discussion and consensus building • Shares "air time" and takes care not to interrupt or cut off others		In addition, • Thoroughly prepares for conversations having read and researched the topic • Invites and encourages other speakers to contribute • Shows appreciation for positive and constructive feedback.
COMMITMENT TO SHARED	• Can not describe what constitutes success in the context of the team's task • Impedes teams progress by failing to completes individual tasks on time and with sufficient quality • Provides no positive feedback or unhelpful negative feedback • Devotes less time and effort required to ensure team benchmarks and due dates are met		• Can generally describe what constitutes success in the context of the team's task • Completes individual tasks on time and with sufficient quality so, but needs some prodding and reminding • Provides intermittent constructive feedback to team members • Devotes the time and effort required to ensure team benchmarks and due dates are met		• Can clearly and specifically describe what constitutes success in the context of the team's task • Completes individual tasks on time and with sufficient quality so as not to impede the team progress toward success • Provides positive and constructive feedback to team members • Devotes the time and effort required to ensure team benchmarks and due dates are met and that work is done to a high standard • Supports others to complete necessary work and ensure the team's success		In addition, • Works to make sure everyone knows what needs to be done • Actively encourages and motivates others to attain high levels of achievement
TEAM & LEADERSHIP	• Has difficulty describing the short and long-term tasks of the team's work • Does not monitor individual or team progress and must repeatedly be given direction • Has difficulty describing the roles and responsibilities of each team member • Has difficulty taking direction from others		• Can generally describe the short and long term tasks of the team's work with some confusion • Monitors individual progress but is less aware of team needs and next steps • Can generally describe what roles and responsibilities each member of the team is expected to perform • Can effectively take direction from others, but does not play a leadership role		• Can clearly and specifically describe the short and long term tasks of the team's work • Monitors progress of team's efforts and is aware of team needs and next steps • Can clearly and specifically describe what roles and responsibilities each member of the team is expected to perform and how they are connected • Can effectively play leadership roles by managing others, but can also take direction from others		In addition, • Works to ensure all team members understand the short and long term tasks • Provides helpful feedback to team on progress • Selects and leverages the most applicable protocols or processes for team management

Created with support from Stanford Center for Assessment, Learning, and Equity (SCALE) and based on similar rubrics from Envision Schools.

©Copyright New Tech Network 2016

Fig. 20.2 NTN Collaboration Rubric

Academies were NTN schools. In 2022, the NTN school model was recognized as the only vetted 'whole school model' by the Texas Education Agency (TEA, 2022).

Integrating Communication and Collaboration Into Science and Technology Education

As can be seen from the previous section, communication is a skill that is used in academic, professional and personal settings. In classrooms, communication can be grouped into three major categories: (1) Verbal communication; (2) Written communication; and (3)Non-verbal communication.

Verbal communication requires the ability to express oneself and, more importantly, to listen to others, especially in a collaborative setting. In the science and technology field, verbal communication can include presentations, videos, scientific posters and online platforms. Written communication requires the ability to choose appropriate academic vocabulary and understand the meaning of academic language. Non-verbal communication requires the ability to observe another person and infer their meeting. In the collaboration context, this requires the additional skill of asking good questions for clarification and providing critical feedback to assure group success.

In science and technology, we are often trying to understand complex systems or trying to solve complex problems. Collaboration in science and technology requires co-operation and co-ordination of effort to achieve a common goal. What makes this possible is good communication.

Collaboration in the Science and Technology Classroom

We know that PBL is an instructional strategy that promotes both collaboration and communication. That said, PBL is not always a viable option for all instructional settings. There is a large literature base on collaborative learning, which can provide specific strategies that can be implemented in classrooms. Probably best known is co-operative learning approaches, which are grounded in social learning theory (Vygotsky, 1962) and social interdependence theory (Johnson, 2003). Co-operative learning is the foundation for most active learning pedagogies, including PBL.

There are a number of practices to facilitate collaboration in the science and technology classroom. At their core is the establishment of students working together in groups. Students can be grouped in any number of configurations, but there are some considerations. Groups should be small enough to allow all members to contribute and large enough to assure diversity of thought. There is no magic number, but three to four group members is a good guide for educators. These groups are small enough to allow all individuals to fully participate. Groups should also develop goals. Providing students with checklists and rubrics can help students to create shared goals when working together. There should also be a set of group operating rules or norms to assure interpersonal communication. It may be a good idea to assign roles to students working in collaborative groups and rotate these roles on a regular basis. Roles let students know what is expected of them in the group, and also mirror how many science and technology organizations operate in government and industry.

Whether using PBL or another pedagogical approach, it is good practice to provide students with a complex problem to solve that is based on a real-world science issue. This can include the creation of scenarios that include engineering design, which require students to apply science concepts and technologies to address the scenario. Well-developed scenarios should allow for varied interpretation that allows for some creativity, another twenty-first century learning skill rather than all group products being identical.

Where possible, students should have the opportunity to utilize online collaboration platforms. This is becoming more common due to the COVID-19 pandemic. Collaboration can be very effective using platforms such as Zoom, Padlet, Google Groups, etc. Many schools have learning management systems that can be used for online collaboration.

Collaborative learning requires effective communication. It is essential that, when implementing collaborative learning in science and technology, students are provided with learning tasks that foster effective verbal and written communication. One way to begin modeling effective communication is to have students attend science and technology conference sessions as observers at first, and then as participants. Educators can utilize videos of model presentations, both oral and written, including multimedia to provide examples and set expectations. There are a number of venues such as the GLOBE International Virtual Science Symposium or the Science Journal for Kids that maintain archives of presentations and written journal-style academic articles.

Science and technology educators can also provide students with scaffolding tools, such as scientific poster templates and PowerPoint presentation templates, to help students learn the elements of effective presentations. Figure 20.3 shows a

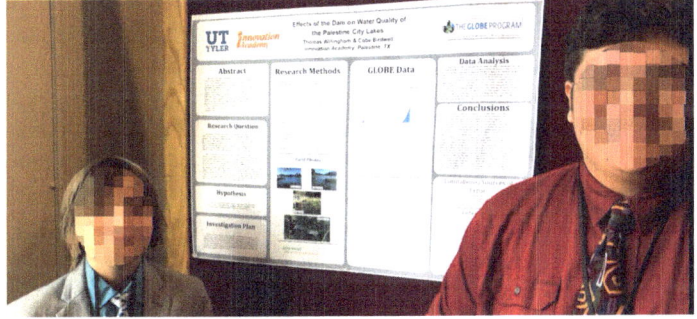

Fig. 20.3 Student science poster

student research poster that was presented at a student science and technology conference. Giving students the opportunity to experience the conventions of a poster session exposes them to the social practices of professional science and technology education academic culture (Mayfield et al., 2018). The students presented as a science professional would present their research findings at a professional venue. This provides students with a valuable real-world workforce experience.

Providing students with instruction on how to communicate science by participating in the development of group products such as manuscripts, posters and scripted multimedia projects can have other positive results, such as deeper understanding of the content that they are learning. Other benefits may include improvement in listening skills, writing skills and oral skills in general. Another benefit of collaborative projects is the possible reduction of individual pressure on one student by providing a peer support system beyond the classroom teacher.

Examples from Primary, Secondary and Tertiary Education

The GLOBE Program (www.globe.gov) is an international science program available to students, educators and scientists, at no cost, in 126 countries. The program is available to all students in primary, secondary and tertiary education. The GLOBE program provides a platform that can support students' development of science and technology communication and collaboration skills through authentic science. GLOBE provides students with tools and protocols with which to collect environmental data and offers a platform to share those data, including online conferences and publications. GLOBE also connects students, educators and scientists with real-world science and satellite missions. Students can collaborate on projects within their school, region, country or worldwide. Students are also provided with the opportunity to communicate their findings through multiple scientific venues, in-person and virtual, so that they can experience science as a professional scientist.

Students can develop collaborative research projects around a local or global problem. They collect local data and GLOBE provides online tools to visualize those data, as well as link their data to global data sets. Figure 20.4 shows a visualization sample. Having online tools was especially important during the height of the COVID-19 pandemic. The system allowed for students and teachers to work on real-world science projects remotely, using technology as the interface.

After completing their data collection, students have multiple venues in which to communicate their results. They can produce a manuscript or a scientific poster, which can be uploaded to the GLOBE website for scientific review, feedback and eventual publication. The GLOBE Student Research Report site provides teachers and students with resources to facilitate the preparation and submission process. Figure 20.5 provides a snapshot of the GLOBE Student Research Report website.

Students can also present their products at the International Virtual Science Symposium, or at the GLOBE Learning Expedition (GLE). GLEs are international student science symposia held every few years. The conferences provide students

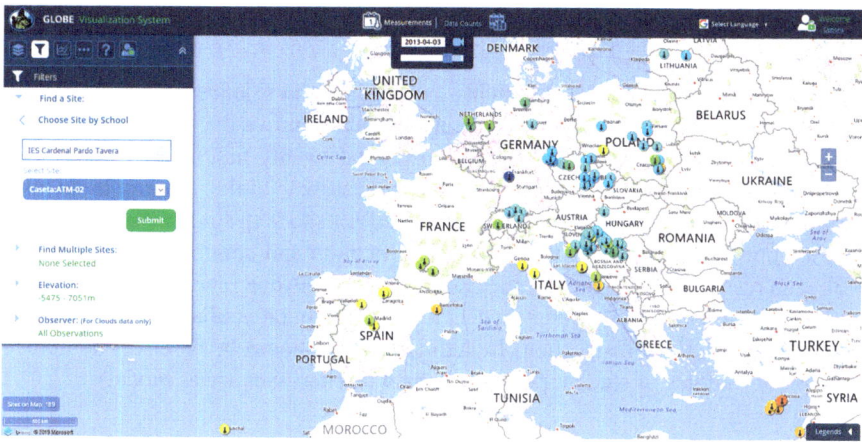

Fig. 20.4 GLOBE visualization platform

Fig. 20.5 GLOBE student research reports

with a chance to present their research and develop international collaborations similar to those of professional scientists. GLEs have been held in the US, Europe, Africa and Asia. Educators also attend the GLEs to share ideas around collaboration and the communication of science.

Discussion and Recommendations

Providing students with the opportunity to participate in science and technology in the same way as science professionals can not only improve the learning of science content, but also help students to develop critical twenty-first century skills. The global COVID-19 pandemic provided many educators with the opportunity to pilot virtual collaboration and learning tools and platforms to help students learn science. The adoption of pedagogies that foster collaboration and communication, such as PBL, can result in positive academic and post-academic outcomes. Further research should focus on how to prepare teachers to best incorporate collaboration and communication into their instruction. In addition, there should be an effort to archive exemplary samples of student products, which teachers can use as models.

Summary

In this chapter, we have discussed that communication and collaboration are two twenty-first century skills that can transform teaching and learning. Communication and collaboration can be effectively taught in the science and technology classroom to better engage students in learning and develop skills needed in their personal and professional lives. Project-based learning and problem-based learning are two instructional approaches that can facilitate the development of communication and collaboration. Providing students with templates for scientific communication can help to scaffold science and technology learning.

Recommended Resources

P21 Network: twenty-first Century Skills https://www.battelleforkids.org/networks/p21
New Tech Network Research https://newtechnetwork.org/impact/
Science Journal for Kids https://www.sciencejournalforkids.org/
GLOBE Student Research Reports Archive https://www.globe.gov/do-globe/research-resources/student-research-reports?p_p_id=gov_globe_cms_projects_ProjectsWebPortlet&_gov_globe_cms_projects_ProjectsWebPortlet_titleFilter=&_gov_globe_cms_projects_ProjectsWebPortlet_schoolNameFilter=&_gov_globe_cms_projects_ProjectsWebPortlet_articleIdFilter=&_gov_globe_cms_projects_ProjectsWebPortlet_reportTypes=&_gov_globe_cms_projects_ProjectsWebPortlet_yearFilter=0&_gov_globe_cms_projects_ProjectsWebPortlet_orgFilterId=0&_gov_globe_cms_projects_ProjectsWebPortlet_gradeLevel=&_gov_globe_cms_projects_ProjectsWebPortlet_collegeCategory=&_gov_globe_cms_projects_

ProjectsWebPortlet_protocolIds=&_gov_globe_cms_projects_
ProjectsWebPortlet_sortCol=4&_gov_globe_cms_projects_
ProjectsWebPortlet_displayStart=0
GLOBE International Virtual Science Symposium https://www.globe.gov/news-events/globe-events/virtual-conferences

References

Aurbach, E. L., Prater, K. E., Cloyd, E. T., & Lindenfeld, L. (2019). *Foundational skills for science communication: A preliminary framework* [white paper]. Retrieved from the University of Michigan: http://hdl.handle.net/2027.42/150489. https://doi.org/10.3998/2027.42/150489.

Barell, J. (2010). Problem-based learning: The foundation for 21st century skills. In J. Bellanca & R. Brandt (Eds.), *21st century skills: Rethinking how students learn*. Solution Tree.

Brownell, S. E., Price, J. V. & Steinman, L. (2013, October 15). Science communication to the general public: Why we need to teach undergraduate and graduate students this skill as part of their formal scientific training. *Journal of Undergraduate Neuroscience Education, 12*(1), E6–10.

Gascoigne, T., & Schiele, B. (1995). In T. Gascoigne, B. Schiele, J. Leach, M. Reidliner, B. Lewenstein, L. Massarani, & P. Broks (Eds.), *Communicating science: A global perspective*. ANU Press Acton ACT 2601.

Johnson, D. W. (2003). Social interdependence: Interrelationships among theory, Research, and practice. *American Psychologist, 58*(11), 934–945. https://doi.org/10.1037/0003-066X.58.11.934

Mayfield, T. J., Olimpo, J. T., Floyd, K. W., & Greenbaum, E. (2018). Collaborative posters to develop students' ability to communicate undervalued scientific resources to non-scientists. *Journal of Microbiology and Biology Education, 19*(1) ASM Journals.

New Tech Network. (2016a). *Collaboration Rubric*. Retrieved from: https://newtechnetwork.org/resources/new-tech-network-collaboration-rubrics/

New Tech Network. (2016b). *Oral Collaboration Rubric*. Retrieved from: https://newtechnetwork.org/resources/ntn-oral-communication-rubrics/

New Tech Network. (2016c). *Written Communication Rubric*. Retrieved from: https://newtechnetwork.org/resources/written-communication-rubrics/

Owens, A. D., & Hite, R. L. (2022). Enhancing student communication competencies in STEM using virtual global collaboration project-based learning. *Research in Science & Technological Education, 40*(1), 76–102.

Partnership for 21st Century Learning. (2015). *P21 Framework Definitions*. Retrieved from: http://www.p21.org/storage/documents/docs/P21_Framework_Definitions_New_Logo_2015.pdf

Saavedra, A. R., Liu, Y., Haderlein, S. K., Rapaport, A., Garland, M., Hoepfner, D., Morgan, K. L., Hu, A., & Lucas Education Research. (2021). *Project-based learning boosts student achievement in AP courses*. Lucas Education Research.

Texas Education Agency. (2022). *ESF vetted school model*. Retrieved from: https://texasesf.org/vetted-programs/

Vygotsky, L. S. (1962). *Thought and language*. MIT Press. (Original work published in 1934).

Michael R. L. Odell is a Professor of STEM Education and holds the Sam and Celia Roosth Chair in the College of Education and Psychology. He holds appointments in the School of Education and the College of Engineering. He is the Co-Founder of the University Academy Laboratory Schools and serves on the School Board. He also provides oversight for the UA Curriculum. He is the Co-Director of the UTeach STEM Teacher Preparation program and the Co-Director of the

EdD in School Improvement Program. His research interests are education policy, sustainable Education, PBL, school improvement, and STEM Education. Dr. Odell holds a PhD in Curriculum and Instruction from Indiana University.

Kelly Dyer is currently an Instructional Coach at the University of Texas at Tyler University Academy, which is a K-12 University Charter school with a focus in STEM. She is also an adjunct faculty member and Master Teacher for the UTeach STEM Teacher preparation program and the Masters in Instructional Coaching in the School of Education. Her research interests include STEM education, dual credit, and blended learning. Dr. Dyer holds a PhD in Curriculum and Instruction from Texas Tech University.

Mitchell D. Klett is currently Professor of Science Education at the University of Northern Michigan. He teaches science and pedagogy courses for pre-service teachers at the elementary and secondary levels. His research interests include Earth Systems Science Education. He earned a PhD in Education with a focus on Curriculum and Instruction from the University of Idaho.

Chapter 21
Afterword: Frameworks for Teaching Science and Technology

Ben Akpan

Abstract This chapter examines four teaching frameworks that have emerged from this book. These are: Philosophy, nature of science and technology as a teaching framework; Teaching science and technology with the curriculum as focus; Teaching science and technology in the context of society; and Teaching science and technology for cross-cutting skills. The chapter ends with concluding remarks and a summary.

Keywords Cross-cutting skills · Science, technology and society · Philosophy and nature of science and technology · And curriculum design, assessment and evaluation

Introduction

In this final chapter of the book, *Contemporary Issues in Science and Technology Education,* I have taken a look at the book in its entirety from the perspective of science and technology teaching and learning, and have identified some ideas and trends in the form of frameworks for teaching. These have been grouped into the following four frameworks: (1) Philosophy, nature of science and technology as a teaching framework – encompassing the ideas in Chaps. 1, 2, 3, 15 and 17; (2) Teaching science and technology with curriculum as focus – considering curriculum design, assessment and evaluation in Chaps. 5 and 6, mathematics and language in the service of science and technology education in Chaps. 7 and 8, STEM as a teaching strategy in Chaps. 4 and 15, and the real and virtual laboratories in Chaps. 9 and 15; (3) Teaching science and technology in the context of society – comprising the ideas in sustainable development in Chap. 10, public understanding of science in Chap. 13, interpreting everyday science in Chap. 11, and indigenous

B. Akpan (✉)
The STAN Place, Abuja, Nigeria

B. Akpan et al. (eds.), *Contemporary Issues in Science and Technology Education*, Contemporary Trends and Issues in Science Education 56,
https://doi.org/10.1007/978-3-031-24259-5_21

knowledge in Chap. 12; and (4) Teaching science and technology for cross-cutting skills – with ideas from problem-solving in Chaps. 18 and 15, creativity in Chap. 19, collaboration and communication in Chap. 20, and additional ideas in Chap. 14. I will next discuss each framework in detail.

Philosophy, Nature of Science and Technology as a Teaching Framework

In Chap. 1, we learnt that philosophy is concerned about how we come to know about phenomena as well as the process by which science and technology gather information. It was also stated that, in science and technology, explanations are made in mechanistic terms. The mechanistic view, as opposed to teleological view, explains phenomena in terms of the cause by which they arise rather than of the purpose that they serve. The mechanistic view of science, for example, underlies the theory of biological evolution (discussed in Chap. 3) as it asserts, among others, that all living things consist of a unique combination of chemicals organised in unique ways and that variations occur in species even as no two individuals of species are alike. The discussion in Chap. 2, which maintains that understanding the nature of science (NOS) and nature of technology (NOT) requires meaningful learning as espoused by David Ausubel, furthers the case for a teaching framework based on the philosophy and nature of science. In the same vein, Chap. 15 makes a case for the use of philosophy as a teaching method for the development of critical and creative thinking in children. Indeed, the entirety of Chap. 17 is devoted to how to promote the mechanistic reasoning of learners in science and technology using emerging technologies. So, arising from this book is a strong case for a teaching approach that is consistent with the philosophy and nature of science and technology.

In our daily lives, we are increasingly experiencing the impact of science and technology. Daily decisions are required in several aspects of human endeavour, which have direct bearings on science and technology. The choice of our meals, drinks, etc. requires not only scientific knowledge, but also illustrates how much confidence we can place in such knowledge. In turn, this leads us to the understanding (or otherwise) of the NOS or NOT, as the case may be. It has been argued, for example, that NOS is at the heart of scientific disciplines and that a teacher who lacks adequate conception of NOS cannot implement the desired instructional activities, because a functional understanding of NOS is a clear prerequisite to achieving the vision of science teaching and learning specified in relevant reform documents across the world. Incidentally, teachers are critical to the proper implementation of this framework. However, not much has been provided to teachers regarding the teaching of NOS to students, since the provision of professional development depends on available financial resources (Lederman et al., 2013). The situation is virtually the same in the case of NOT, because NOT is given little attention in classrooms. Pleasants et al. (2019) have suggested that a great deal of effort has to be put in to achieve the desired effective NOT teaching and learning, because any achievement is predicated on the ability of classroom teachers who are required to be appropriately skilled at delivery of NOT to the students. According to them, responsibility for furthering NOT instruction falls on policymakers in education as well as teacher educators. It would appear, therefore, that faculties and institutes of education that are responsible for pre-science and in-service training of teachers need to do more to promote and further the acquisition of knowledge and understanding of NOS and NOT by the various participants in their training programmes.

Teaching Science and Technology with the Curriculum as Focus

Chapter 5 provides an international perspective on the design of the curriculum in science and technology education. It stresses the need for alignment between objectives (specific statements that indicate the intention of teaching) and learning outcomes (what the teacher expects the learners to know, understand or/and be capable of doing when the learning process is completed). Chapter 6 goes further to examine assessment in two dimensions: the one, *formative*, which sets out to improve performance in the course of the teaching process; the other, *summative*, which is performed after the teaching process has been completed and is aimed towards the outcome. Incidentally, a flurry of ideas has emerged in the book that point to the desirability of teaching science and technology with the curriculum as focus. I will highlight the ideas in the ensuing section.

Mathematics and Language in Science and Technology Teaching and Learning

Chapters 7 and 8 respectively focus on the role of mathematics and language in the teaching and learning of science and technology subjects. With a focus on the curriculum, Chap. 7 signals a dual challenge facing mathematics teachers: (i) how to address the relatively low level of engagement of students in school mathematics; and (ii) how to prepare students for the application of mathematics in current and future studies in science and technology subjects. It advocates moving school mathematics beyond the present conception of being a subject with rules and procedures to an altogether new level where mathematics could provide knowledge and skills for solving science and technology – related issues and problems. Of course, there is a rationale for this recommendation. Literacy in mathematics should be seen within the context of individuals carrying out their daily tasks more effectively and seamlessly as they apply computational, quantitative and spatial reasoning skills. In addition, the application of mathematical concepts and skills is an overarching prerequisite for learning many concepts in electronics, for example. Yet, even more fundamental is the fact that mathematics is a part and parcel of human culture and it behoves us to pass on that heritage. That heritage subsists because mathematics, like science or technology, is a universal language as it has no national boundaries. A rule in mathematics doesn't change as you traverse the globe. Still, mathematics continues to serve as a potent vehicle for interdisciplinary approaches to curriculum development, implementation, and evaluation.

With respect to the importance of language in science and technology teaching and learning, the opinion in Chap. 8 is emphatic: language is the most valuable psychological tool in matters of communication between the learner and the learning environment, to the extent that learners with the same IQ level may show disparities in learning abilities on account of differences in language proficiency and capability. This is why, for instance, PhysicsCatalyst (2022) advocates a *language across the curriculum* approach. According to the organisation, irrespective of the subject that the learners are studying, they assimilate new concepts and ideas through the use of language. In general, therefore, language helps learners to learn content, expand ideas, collect technical terms in different subjects, access various careers and to carry out self-studies. Thus, language is more than communication skills, it is linked to the thinking process, even as it is a tool for making meaning of pieces of information. Additionally, language supports mental activities as well as precision in cognition, and is of overarching importance for learners of all ages to function optimally in academic endeavours.

STEM as a Teaching Strategy

As noted elsewhere (Akpan, 2021a), the overall goal of STEM is to discover and apply knowledge gained from nature in furtherance of human development. Through the exploration and formulation of the general principles of nature, STEM tends to recognise what is common in different patterns, and what may be different in similar situations. STEM, thus, tries to understand our world by studying the mechanisms and interrelations of phenomena through various research efforts. Even so, STEM is not only research, because its larger proportion inevitably relates to development. Yet, research is at the front edge of STEM (*ibid*). Emerging from this book is a furtherance of the role of STEM, to wit: *STEM as a teaching strategy*. Chapter 4 makes a case for STEM education as a meta-discipline, which involves an integration that leads to the breaking of barriers as well as boundaries that exist between the different STEM subjects. The result is the delivery of STEM through an interdisciplinary approach. In general, this approach involves a merger of traditional educational concepts or methods so as to arrive at new approaches or solutions. Holbrook and Rannikmae (2019) contend that interdisciplinarity is not a simple addition of parts, but the recognition that each discipline can affect the research output of another. Indeed, Chap. 15 makes a case for the *STEM model of teaching* – an educational and economic strategy with the aim of facilitating the acquisition of modern-day knowledge and skills. It is sometimes the case that inputs from the Arts and humanities can transform *STEM* education into a transdisciplinary *STEAM* education, the additional 'A' representing the *Arts*. McGregor (2015) is of the view that transdisciplinarity is *beyond* disciplines, with a new knowledge about what is between, across and beyond disciplines (hence the term *trans*).

The Real and Virtual Laboratories

Hugerat and Hofstein (2019) are of the view that laboratory activities have a major role in science curricula, as they provide a means of making sense of natural phenomena as well as ensuring that the teaching and learning of science is not only relevant, but highly motivating. This position is being supported by the ideas emerging from this book. Chapter 15 has made a case for the laboratory as a teaching strategy, even as the traditional laboratories are transitioning into virtual laboratories due, in part, to financial issues as well as health crises as witnessed during the COVID-19 pandemic. Chapter 9 examines the opportunities and challenges in the use of virtual laboratory-based practical work, as well as the likelihood and extent to which virtual laboratories will replace real laboratories in science and technology teaching and learning. Some science and technology teachers suggest that the laboratory assists the learners in comprehending concepts, as well as in the construction of knowledge, because the laboratories ensure that learners integrate and combine both hands-on and minds-on activities at the same time (Hugerat & Hofstein, 2019). The view here is that virtual and real laboratories complement each other.

Teaching Science and Technology in the Context of the Society

According to Holbrook and Rannikmae (2017), context-based teaching occurs when a lesson is introduced from a real-world context by a teacher, who relates it to the learning of specific concepts. This is in contradistinction to teaching science content based on the curriculum. The basis of context-based teaching and learning is to ensure relevance to the learners and to promote motivation for learning. Context-based learning mitigates curriculum overload and allows learners to transfer learning to new situations (*ibid*). Some of the issues emerging from this book point to context-based teaching, specifically: teaching science and technology in the context of society. Four chapters stand out in this regard. Chapter 10 examines the concept of sustainability as it relates to global development. It also considers the need to develop a global workforce that is literate in STEM so as to ensure that people can meet the present and future challenges. This is an interesting context, where the society is the focus.

Chapter 11 presents an interesting and illuminating perspective on the experiences that children have on their first contact with science learning. Appropriately captioned *Interpreting everyday science*, the chapter notes that, in the beginning, young children try to observe as well as make sense of the world. This provides a great moment through which the children can firm up their imaginations if they are in formal school settings. Schools can seize this as an auspicious opportunity to ensure that great foundations are established for the children's science learning. In general, children learn science through *trial and error*. It is great fun. Children require a great deal of time and attention as they try things out, and look for some clues. Children should be given sufficient time and an appropriate learning environment to make sense out of phenomena and to discover, gradually, with guidance. Research into children's ideas indicate that children try to make sense of things around them long before we interfere by *teaching* them (Harlen, 2001). It is also known that, when we do intervene, we need to take the children's own ideas as a starting point. This means encouraging them to test their own ideas and alternative, more specific, ones by gathering evidence and using it to see what makes sense, so that they realise the value of more broadly applicable concepts. But this will only be effective if the children know how to collect and use evidence in relation to phenomena that they are studying; that is, if they have the appropriate process skills. Moreover, we should be helping them to develop bigger ideas. *Big* ideas (*ibid*) are the ones that are applicable to, and link together, a range of phenomena, whereas *small* ideas relate to specific situations or things, but which have limited applications beyond them. Incidentally, the big ideas are often too abstract and too remote from specific applications for children to grasp. It would certainly be unproductive to start with the big ideas, which are unlikely to be understood. Thus, learning has to start from small ideas, which relate to specific events or phenomena that are *close* to the children. To develop more widely applicable, bigger ideas, we have to link together events that, once encountered, can be explained in similar ways. Children bring their own ideas and experiences to classrooms and these experiences shape their knowledge. Teachers need to uncover children's prior knowledge about the topic that will be taught in class. By understanding the prior knowledge of children, teachers can select appropriate instructional strategies to help children build upon the knowledge that they bring to the classroom.

Chapter 13 explores public understanding of science and technology. Again, the context is society-based. The chapter maintains that people must be willing to learn science and technology in order to meet societal challenges. Providing the example of the challenges posed by COVID-19, the chapter is of the view that people need to understand the science behind what affects their personal wellbeing, such as disease prevention, reason for vaccination, need for public health and healthy meals; and online services and programmes. Chapter 12 examines indigenous knowledge and science and technology education. Basically, *indigenous people* are the *first peoples* to live in a particular area of the world. Blades and McIvor (2017) contend that these *nations* have varied political organisation and lifestyle, as well as culture. In addition to the diversity, there are beliefs, orientations and traditional knowledge that may run in conflict with the tenets of mainstream science and technology. Incidentally, learners who have an indigenous inheritance have been known to be persuaded to demonstrate an understanding of the world through the prism of mainstream or *Western* viewpoints. Blades and McIvor stress the need to introduce all learners to indigenous ways of thinking, because such inclusion leads to a socially-just, inclusive pedagogy and presents an opportunity for learners to develop a more expansive understanding of science and technology. Indeed, Chap. 12 shares the same viewpoint. The chapter maintains that indigenous knowledge uses holistic worldviews and integrates physical, spiritual and moral perspectives. It maintains that indigenous knowledge has the potential to open up opportunities for scientists and citizens to contribute to the advancement of science and technology, in addition to seeking solutions to a myriad of challenges such as global crisis (for instance, the COVID-19 pandemic) and environmental degradation, as well as biodiversity loss.

Teaching Science and Technology for Cross-Cutting Skills

Cross-cutting skills are those skills that enhance learning and application in virtually all disciplines and careers. The skills include oral/written communication, critical thinking, working effectively in teams, collaboration, ethics and accountability, creativity, analytical thinking and problem-solving. At workplaces, employers prefer staff members who demonstrate skills that permeate all fields of human endeavour, even though these employers value discipline-specific knowledge. The corollary is that educational institutions should place a high premium on preparing learners to acquire these cross-cutting skills. These skills are also referred to as *twenty-first century skills*. The skills are not discipline-specific, but can be acquired in schools and related settings. Cross-cutting skills are helpful in preparing individuals for the world of work and lifelong learning in a globalised world (Andrade, 2020). This book has highlighted the importance of teaching science and technology for some of these cross-cutting skills: problem-solving, creativity, collaboration and communication.

Chapter 15 asserts that problem-solving is widely used as a strategy to organise didactic units that comprise a collection of carefully selected and sequentially arranged problems. The chapter identifies the distinctive feature of the problem-solving methodology as being the gradual increase in the complexity of the sequenced problems to be solved, and posits that the methodology is anchored on self-regulated learning, since the learners are responsible for their own learning. In the same vein, Chap. 18 discusses problem-solving in science and technology education and is of the opinion that problem-solving involves describing an identified problem, finding out the cause of the problem, identifying/analysing the likely solutions, and finally applying the solutions so identified in meeting the challenges in science and technology. The chapter, relying on the work of Garrett opines that problem-solving can facilitate the acquisition of many metacognitive skills in science and technology education. The link of problem-solving to *metacognition* is very instructive, because it is in the nature of cognitive theory to use problem-solving to explain complex forms of learning with emphasis on transfer of knowledge to the learner in the most effective manner possible. In line with cognitive theory, there is the overarching need for a learner to be able to implement knowledge in different contexts and conditions. When this happens, the learner is said to have the capacity to *transfer knowledge*. Two of the techniques used in assuring effective transfer of knowledge are *simplification* and *standardisation*. This implies that knowledge can be analysed, disaggregated and simplified into basic building blocks. This is not at all possible in the absence of metacognitive skills. My great friend, Emeritus Professor Keith Taber, has elaborated on the concept of metacognition – the awareness and understanding of one's own thought processes (see Chap. 14). Metacognition involves *thinking about one's thinking*. It refers to the processes used to plan, monitor and assess one's understanding and performance. Encouraging students to think about the way in which they do things and to reflect on their own thinking is introducing them to metacognition. When we are conscious of doing something, and reflect on its value, we are more likely to apply that kind of thinking again in a future situation where it is relevant. Hence, it has considerable value for lifelong learning. The involvement of students in assessing their own learning both requires and encourages students to reflect on learning. In order for this to happen, teachers should ensure that assessment (see Chap. 6) is used to adapt teaching; feedback is given to students in terms of how to improve their work, not in terms of judgemental comments; students are actively engaged in self-assessment and in helping to decide on their next steps; and that all students are regarded as being capable of learning.

Chapter 19 of this book examines the concept of creativity in science and technology education. The chapter identifies the content, curricular focus and personal characteristics as factors affecting creativity, while the tools for creativity include digital learning, asynchronous and independent access to digitised information. Basically, creativity refers to the capacity to generate ideas that are useful in solving problems. It involves inventiveness. There are four stages in the creative process: (1) Preparation – a period of a rather intense and conscious attention to the task that continues until there is a need to take a break and refresh the working memory; (2) Incubation – a period of relaxation away from the task in order for the brain to look at the problem from a variety of perspectives; (3) Illumination – this is the *eureka* moment when the solution suddenly becomes clear; and (4) Verification: this involves checking thoroughly to find out if the solution really works (Akpan, 2021b). Quite surprisingly, two ideas emerging from recent research on the creative process seem to go against long-held viewpoints: the more focused we are on solving a task, the more difficult it is to do so; and insights come to us more readily when we are more relaxed (Sousa & Pilecki, 2018).

The preceding chapter in this book discussed two additional cross-cutting skills, namely collaboration and communication. The authors maintain that communication and collaboration are skills that are, by and large, highly interdependent and are very useful in problem-based, as well as project-based, instruction. According to them, these two cross-cutting skills are useful beyond the classroom as they can be, and are currently being, used in personal and professional life. Akpan (2020) is of the view that *collaboration*, in the context of science and technology education, refers to the action of working with another person or in a group to produce something of positive value. An example of collaboration is that of a team of more than 30 surgeons in India in November 2007 who performed a gruelling 24-hour operation on *Lakshmi Tatma* (*ibid*). Lakshmi was born on 31st October 2005 in Araria district, Bihar, India, with four arms and four legs. Scientists say that she was in fact one of a pair of *ischiopagus conjoined twins*, one of whom was without a head as this had wasted away (*atrophied*) and the chest did not fully develop in the womb, resulting in the one surviving child with four arms and four legs. So, the surviving foetus absorbed the limbs, kidneys and some body parts of the underdeveloped foetus. The doctors worked through the night to remove the extra limbs and organs. A team of neurologists separated the fused spines, while orthopaedic surgeons removed most of the 'parasite', carefully identifying which organs and internal structures belonged to the surviving girl. The operation also involved the transplanting of a good kidney from the other twin into Lakshmi.

Collaborative learning is well known (Akpan, 2021b), yet teachers often create instructional environments that result in students working individually. By collaborating, students have opportunities to contemplate the information that they encounter, discuss their emerging ideas, and evaluate their understandings. The back-and-forth exchange between two or more students can encourage a deeper understanding of the content. In science and technology, collaboration allows students to experience the social component that is inherent in all activities. In order to support collaborative learning, students need to be physically close to one another so that they can discuss the presented information. This means gathering in small groups, working together during laboratory sessions, or sitting around a table. In addition, students need guidance in learning how to work in a collaborative way. The teacher can present guidelines that support collaborative conversations, which can include: only one person talking at a time, an acknowledgement of the ideas presented, and every person endeavouring to participate in the conversation. These guidelines help to ensure that all students have an opportunity to participate in the conversation.

With respect to communication, learners communicate by articulating their thoughts and ideas through oral, written and/or non-verbal means. It involves the use of multimedia formats such as video and imagery. The characteristics of effective communication include: (1) Clarity – using concrete and precise language to get points across; (2) Conciseness – getting straight to the point by avoiding wordiness, empty phrases and redundancies; (3) Correctness – ensuring that statements are facts-based and provable, and ensuring that the content is reviewed before sending; (4) Completeness – giving the whole picture; (5) Coherence – ensuring a logical arrangement of points, and compartmentalising if there are multiple points in a single message; (6) Consideration – weighing your words and thinking of the likely effect on the recipient of the message; (7) Courtesy – conveying messages constructively and with respect; and (8) Concreteness – mitigating the risk of misunderstanding by making sure that the message is tangible and supported by facts.

Concluding Remarks

The effective implementation of the frameworks described above depends on the teacher. Indeed, no matter how well and carefully a science and technology programme or curriculum has been constructed, its quality as adequate and relevant depends on the teacher. It is the teachers who, in the final analysis as they work with learners, determine which of the proposed experiences in the curriculum the learners should acquire. This is not a simple matter. In translating a planned curriculum, teachers face two major issues. They must ascertain not only that the experiences that they encourage learners to acquire closely correspond to the intended curriculum experiences, but also that the set of experiences that the learners acquire is consistent with their background, interest and motivation. Thus, in science and technology, how we teach is of special importance because it is in the way that we teach that we convey some of the most important dimensions of the subject matter. The most important elements of science and technology can be communicated only by the teacher. We want to stimulate the curiosity of the learners and nourish their desire to find out. We want children to learn how to investigate and design experiments so that they can learn from their experiences. In the beginning, at least, this requires a teacher who can ask the critical questions, demonstrate alternatives, and help when the obstacles seem insurmountable. Certainly, science and technology involve the willingness to consider questions and observations even though they may be inconvenient and upsetting. It is difficult to see how this attitude of open-mindedness can be communicated other than through a teacher model who demonstrates this approach to the children. Yet, in any science and technology classroom, the personality of the teacher is a very limiting factor in teaching processes and strategies. Teachers' temperaments, convictions about human nature and what they consider to be the purpose of science and technology education determine the classroom methods. Indeed, from whatever perspective in which curriculum implementation in science and technology is viewed, the overall goal is towards the acquisition of scientific and technological literacy (STL) – the capacity to apply scientific and technological knowledge in making evidence-based decisions about the natural world and the changes made to it by humans. A consideration of STL has implications for *how* we teach and *what* we teach (see pedagogical content knowledge in Chap. 16). Harlen (2001) gives two reasons for this. First, although the acquisition of specific knowledge of a subject is important, the ability to apply what has been learned in everyday life depends crucially on the understanding of broader concepts and applicable mental skills. Second, with the amount of knowledge expanding more and more rapidly, schools cannot provide all the knowledge that people will need in later life, so it is necessary to prepare future citizens for lifelong learning, which means *learning how to learn* and, perhaps more importantly, *enjoying learning*. In all these, the teacher is the key!

Summary

In this chapter, I have identified and discussed four frameworks that have emerged from the book to which the teaching and learning of science and technology may be anchored. These are: Philosophy, nature of science and technology as a teaching framework; Teaching science and technology with the curriculum as focus; Teaching science and technology in the context of society; and Teaching science and technology for cross-cutting skills. I have opined that a clear knowledge of the *nature of science* along with the *nature of technology* by teachers is a prerequisite to achieving the vision of science and technology teaching and learning that is in line with current curriculum reform documents across the world. I have maintained that mathematics and language are great *tools* for the teaching and learning of science and technology, just as STEM, sometimes viewed solely through the prism of research and development, is gradually gaining attention as a teaching strategy that can transition from an *interdisciplinary* (STEM) approach to a *transdisciplinary* (STEAM) methodology. The chapter notes that *virtual* laboratories are complementing the role of the traditional (*real*) laboratories, and recommends the teaching of science and technology in the context of society using such topics as sustainability, early childhood science, and indigenous knowledge. Additionally, the chapter advocates the teaching of science and technology for cross-cutting skills, such as problem-solving, creativity, collaboration and communication. In the end, I have emphasised the overarching pivotal *role of the teacher* in the teaching and learning of science and technology.

Recommended Resources

Best resources for science teachers https://www.educatorstechnology.com/2014/02/the-25-must-have-resources-for-science.html

Science and technology teacher resources https://americanart.si.edu/education/k-12/resources/science

Science resources and technology tools for teachers https://edtechteacher.org/tools/science/

Teaching resources for the science classroom https://www.calacademy.org/educators/teaching-resources-for-the-science-classroom

References

Akpan, B. (2020). Towards a higher education programme in STEM that facilitates the acquisition of 21st century skills. In S. Akinrinade, S. Oyewweso, S. G. Odewumi, & A. Kola-Olusanya (Eds.), *Pivotal issues in higher education development in Nigeria – Essays in honour of distinguished Professor Peter Okebukola,OFR* (pp. 417–429). University Press Plc.

Akpan, B. (2021a). *Ethics and STEM education.* Keynote Address to the 61st Annual (Virtual) Conference of the Science Teachers Association of Nigeria (STAN), 23–26 August.

Akpan, B. (2021b). *STEAM and collaborative education.* Keynote address to the 35th Annual Congress of the Nigerian Academy of Education, University of Abuja, 9th November.

Andrade, M. S. (2020). Cross-cutting skills: Strategies for teaching and learning. *Higher Education Pedagogies, 5*(1), 165–181. Retrieved from: https://doi.org/10.1080/23752696.2020.1810096. Accessed 12 November 2022

Blades, D., & McIvor, O. (2017). Science education and indigenous learners. In K. S. Taber & B. Akpan (Eds.), *Science education: An international course companion* (pp. 465–478). Sense Publishers.

Harlen, W. (2001). *Primary science: Taking the plunge* (2nd ed.). Heinemann.

Holbrook, J., & Rannikmae, M. (2017). Context-based teaching and socio-scientific issues. In K. S. Taber & B. Akpan (Eds.), *Science education: An international course companion* (pp. 290–305). Sense Publishers.

Holbrook, J., & Rannikmae, M. (2019). Interdisciplinarity and transdisciplinarity. In B. Akpan (Ed.), *Science education: Visions of the future* (pp. 79–90). Next Generation Education.

Hugerat, M., & Hofstein, A. (2019). The future of science laboratories. In B. Akpan (Ed.), *Science education: Visions of the future* (pp. 197–214). Next Generation Education.

Lederman, N. G., Lederman, J. S., & Antink, A. (2013). Nature of science and scientific inquiry as contexts for the learning of science and achievement of scientific literacy. *International Journal of Education in Mathematics, Science and Technology, I*(3), 138–147.

McGregor, S. L. T. (2015). Transdisciplinary knowledge creation. In P. T. Gibbs (Ed.), *Transdisciplinary professional knowledge and practice* (p. pps. 9–24). Springer.

PhysicsCatalyst (2022) *Language across the curriculum approach.* Retrieved from: https://physics-catalyst.com/graduation/language-across-curriculum-approach/. Accessed 11 November 2022.

Pleasants, J., Clough, M. P., & Olson, J. K. (2019). The urgent need to address the nature of technology: Implications for science education. In B. Akpan (Ed.), *Science education: Visions of the future* (pp. 31–46). Next Generation Education.

Sousa, D. A., & Pilecki, T. (2018). *From STEM to STEAM: Brain-compatible strategies and lessons that integrate the arts* (2nd ed.). Corwin.

Ben Akpan, a professor of science education, is the Executive Director of the Science Teachers Association of Nigeria (STAN). He served as President of the International Council of Associations for Science Education (ICASE) for 2011–2013 and currently serves on the Executive Committee of ICASE as the Chair of the World Conferences Standing Committee. Ben's areas of interest include chemistry, science education, environmental education and support for science teacher associations. He is the editor of *Science Education: A Global Perspective,* published by Springer; co-editor (with Keith S. Taber) of *Science Education: An International Course Companion,* published by Sense Publishers; co-editor with Professor Teresa Kennedy of *Science Education in Theory and Practice* published by Springer; and the editor of *Science Education: Visions of the Future,* published by Next Generation Education. Ben is a member of the Editorial Boards of the *Australian Journal of Science and Technology* (AJST), *Journal of Contemporary Educational Research* (JCER), *Action Research and Innovation in Science Education* (ARISE) Journal, and APEduC *Journal on Research and Practices in Science Education, Mathematics, and Technology* – an electronic scientific-didactic publication of the Portuguese Association of Science Education. He is the recipient of many commendations, prizes, and awards.

Index

Printed by Printforce, United Kingdom